The

MARSHES

of Southwestern Lake Erie

Dedication of the Magee Marsh Bird Trail, a mile-long boardwalk that winds through the fragile beach ridge habitat at Magee Marsh Wildlife Area. *Left to right:* Frank McConoughey, Laurel Van Camp, Joe Sommer, Lou Campbell, and Richard Pierce. *Credit: Division of Wildlife.*

The
MARSHES
of Southwestern Lake Erie

Louis W. Campbell
with Claire Gavin

With a Foreword by Harold Mayfield

Ohio University Press
Athens

Ohio University Press, Athens, Ohio 45701
© 1995 by Louis W. Campbell

99 98 97 96 95   5 4 3 2

Ohio University Press books are printed on acid-free paper ∞

Library of Congress Cataloging-in-Publication Data

Campbell, Louis W.
     The marshes of southwestern Lake Erie / Louis W. Campbell
with Claire Gavin ; foreword by Harold Mayfield.
          p.   cm.
     Includes bibliographical references (p.   ) and index.
     ISBN 0-8214-1107-1 (cl)   ISBN 0-8214-1094-6 (pbk)
     1. Marsh ecology—Erie, Lake.   2. Marshes—Erie, Lake.
3. Marsh fauna—Erie, Lake.   I. Gavin, Claire.   II. Title.
QH104.5.E73C36   1993   94-21052
574.5′26325′0977112—dc20   CIP

To my wife, Helen

High in a maple tree, a mocking bird
Serenades with myriad sparkling notes.
The sweetest music he has ever heard
He now purloins from scores of feathered throats,
And to his lady on her nest it floats:

Jewels of melody so pure and true,
With depth and color like the rarest stone.
A robber chieftain, this, who now would sue
His lady's heart with stolen gems of tone,
Flung to the world—but meant for her alone.

—Louis W. Campbell

The Gift of the Glacier   90
The Creation of the Marshes   93
Settling the Marsh Borders   105
Curbing the Marshes   111

Chapter 2: Who Has Owned the Marshes?   117

Cedar Point National Wildlife Refuge   118
Ottawa National Wildlife Refuge   129
    Navarre Marsh   136
    Darby Marsh   139
    Pintail Marsh   139
Magee Marsh Wildlife Area   140
Metzger Marsh Wildlife Area   147
Maumee Bay State Park   148
Sand Beach Marsh   149
Toussaint Marsh   149

Chapter 3: Birds of the Marshes, 1870–1991   154

Bald Eagles   158
Herons   162
Cormorants   166
Gulls and Terns   167
Waterfowl   171
Shorebirds   175

Appendix A: Mammals, Reptiles & Amphibians,
    and Plants of the Marshes   179

Appendix B: Birds of the Ohio Lake Erie Marshes   193

Notes   215

Bibliography   221

Index   225

Acknowledgments

Many, many people over the years have contributed to my knowledge and appreciation of the Lake Erie marshes. I am grateful to Laurel Van Camp, Harold Mayfield, Mark Shieldcastle, Guy Denny, Jack Weeks, Gildo Tori, Karl Bednarik, Michael Tansey, R. L. Meeks, Jeanne Hawkins, Joseph H. Thompson, and Doris Stifel. Many friends and colleagues are now deceased: Milton Trautman, Edward Lamb, Josephine Fassett, Robert Crofts, Lou Klewer, Beatrice Waterbury, Cornelius Mominee, Arthur Secor, Alfred Manke, Nevin O. Winter.

Thanks to Toledo area birders who have contributed their findings to the permanent record: John Anderson, Tom Bartlett, Chris Crofts, Cheryl and Edward Pierce, Paula Jack, Thomas Kemp, Matt Anderson, John Szanto, Dr. Elliott Tramer, Neil Waterbury, Sandra Zenser, Joseph Komorowski, John Redman, Gwen Fick, John Stophlet, Kathy Mock, and Dr. William Smith.

Thanks to all who contributed photographs for this book, including Tom Mominee, John Arnold, the staff at Magee Marsh, and others who are acknowledged throughout the book. Thanks to officers of the Toledo Edison Company for originally suggesting the project.

Thanks to Craig Brunner who redrew my maps of the Lake Erie marshes for this publication.

For critical readings of the manuscript and many helpful suggestions, I am especially grateful to Harold Mayfield, Mark and Julie Shieldcastle (Magee Marsh), Charles Blair and Charles Marshall (Ottawa National Wildlife Refuge), James Brunner, and Robert

Koch. Thanks to Professor Jane Forsyth and to Charles E. Herdendorf, who enriched the book with their knowledge of Great Lakes geology. Thanks to Dr. Michael Schlosser, to Greg Miller, of the Toledo Public Library, and to Isabel Kaplan, of Carlson Library, University of Rochester. Thanks to my daughter, Claire, who more than anyone else was responsible for the publication of this book. And special thanks to my wife, Helen, who has typed many hundreds of pages, encouraged my work, and exhorted me to excellence through more than fifty years of marriage.

Ohio University Press joins the author in thanking the following individuals and organizations who helped the Press support the costs of color illustrations:

Louise N. Bankey; Fred & Rosemary Blodgett; Earl & Thurid Campbell; Dennis M. Forsythe; Lorene Gunderson; Herbert Hover; Paula G. Jack; Richard & Joan Kimple; Carolyn Manchester; Harold Mayfield; Ann Mumford; Ed & Cheryl Pierce; Fred & Doris Plassman; Adele Shelton; Don & Doris Van Buskirk; Daniel Wilkins; Charles & Judy Wilson; Hubert & Norene Zernickow; Black Swamp Bird Observatory; Erie Shores Birding Association; Green Creek Wildlife Society; Maumee Valley Audubon Society; Naturalists' Camera Club of Toledo; *Ohio Cardinal* and Toledo Naturalists' Association.

This project was supported by the Division of Wildlife with funds donated through the DO SOMETHING WILD! state income tax check-off.

Foreword

By Harold F. Mayfield

As our land becomes increasingly crowded, and even the farm woodlots give way to suburbia, many city dwellers become hungry for an escape back to the natural environment. People travel hundreds of miles to vacation where they can enjoy a touch of wilderness. Some remote trails in the national parks and forests are so busy that visitors have to make reservations to use them.

Few people in northwestern Ohio realize that a remnant of the original wilderness lies almost at their doorstep, in the large marshes near the shore of Lake Erie. Here it is possible to find solitude and to move silently beside waterways and cattail-bordered ponds that have changed little since the Indians canoed them centuries ago. Many of them have retained their pristine condition because the wildlife preserves and duck hunting clubs have discouraged visitors.

The rare individuals who know the marshes are mainly a small set of men who have earned a living on the fringe of Lake Erie. They have guided hunters, trapped muskrats for the market, and managed the wetlands. Most of them are of French Canadian descent and have carried down the lore of the marshes from their forebears, who settled near the marshes generations ago.

If there is any outsider who knows the area as well as the marsh people it is Lou Campbell. Indeed, he knows the marshes in greater depth, because he brings to the subject a scientific mind. For more than sixty years he has prowled the marshes with the primary aim of studying their birdlife. As a harmless bird watcher he was accepted by the custodians where the public generally was excluded.

Lou Campbell started as a fisherman, but his love for nature led to a focus on birds, and for decades he has been the acknowledged authority on the birds of northwestern Ohio. Since nature is all of one piece, he is also thoroughly informed on the reptiles, amphibians, insects, mammals, and plants of this region. He has published extensively on all these natural features, and his *Birds of the Toledo Area* (1968), succeeding his *Birds of Lucas County* (1940), has been recognized as a model for reports on the birds of a restricted region. For many years he wrote an outdoor column for the old *Toledo Times* newspaper, through which he maintained close contacts with hunters and fishermen. Lou played a principal role in awakening public interest in the unique Oak Openings area west of Toledo, which led to the establishment of a series of parks and preserves, one of which was named by the state of Ohio in his honor.

Lou Campbell brings to this topic not only a scientific understanding, but also a genuine love of the wild marshlands. The result is a unique work on the charms of a neglected piece of our natural landscape.

Author's Note

In this book I have discussed in depth only the major Lake Erie marshes between Toledo and Port Clinton, Ohio. Other important marshes include the Erie Marsh in Michigan and the marshes of Sandusky Bay: the Winous Point Marsh and the Ottawa Shooting Club Marsh. My decision to limit the scope of the book was based on a number of considerations. First, several other surveys exist for these marshes, and my personal acquaintance with them is limited. Second, the Sandusky Bay marshes are widely separated and not open to the public without permission, whereas those in this book are almost continuous with each other, most are open to the public, and all are located on a major bird migration route. Finally, these marshes, virtually unknown even to area residents, are of great historical as well as ecological interest.

Introduction

Why Preserve the Marshes?

What is a marsh? In the minds of far too many people, it is a foul-smelling patch of stagnant water, spotted with green scum and humming with mosquitoes. Cruising by on paved highways, they gaze out at a sea of waving cattails and shudder, thinking of snakes, insects, malaria, and mud. Why, they wonder, couldn't this wasteland be put to better use? Why not turn it into marinas, farms, vacation homes, or neat parks with grassy playgrounds and paved tennis courts? Many marshes along Lake Erie underwent such changes in the past; now both federal and state governments are committed to protecting and preserving these wetlands. Why?

A freshwater marsh is an open expanse of cattails, grasses, and cane, cut by streams and punctuated with pools of open water. However, the entire marsh complex, or ecosystem, is much more. At the borders of the ponds, cattails, and cane are dense thickets of young trees and shrubs, patches of marsh grass forming wet prairies, and remnants of ancient swamp forests. Between the marshes and Lake Erie are ridges of sand. Winding around and through most of the wetlands, enclosing and protecting them, are reinforced earthen embankments, or dikes. Each of these areas—eight in all—is a separate world, a distinct community of plant and animal life. Because of this rich variety of habitat, the Lake Erie marshes shelter more birds, more mammals, more snakes (none venomous), and more amphibians per acre than any other area in Ohio.

For thousands of years, virtually the entire southwestern shore of Lake Erie was bordered by marshland and swamp forest; the

1

remaining marshes, some thirty thousand acres, are the most primitive lands in Ohio. Deep in the interior of these ancient wetlands, protected from intrusion by law or by sheer inaccessibility, life goes on today much as it did hundreds of years ago. In spring and fall, the marshes are way stations, offering shelter and food to migrating waterfowl, shorebirds, hawks, and multitudes of songbirds on their long journeys to and from their northern breeding grounds. In summer, they are havens for flightless molting ducks. To the great blue herons, black-crowned night herons, and great egrets that nest on West Sister Island, the marshes are a made-to-order cafeteria, nine miles from home—in fact, if the marshes disappeared, the heronry would disappear.

The marshes can sustain all this life and much, much more because of their wealth of habitats:

- **Open ponds**, one to four feet deep, with many bottom-rooted and free-floating aquatic plants. Some of the flowering plants that grow in and around these ponds are among the most beautiful in all of nature. In addition to a few fish and unimaginable numbers of microscopic creatures and insect larvae, the ponds are home to muskrats, many kinds of turtles, water and fox snakes, and leopard frogs and bullfrogs.

- Nearly solid growths of **broad-leafed cattail** up to seven feet high, cut irregularly by open waterways. Wildlife is limited here, though marsh wrens and red-winged blackbirds nest deep inside, and bitterns and coots can be found at the edges. Narrow-leafed cattail is normally found in somewhat drier areas at marsh borders and in shallow ditches.

- **Bluejoint grass meadows**, with occasional patches of slough and cut grass, usually growing in less than six inches of water. Few other plants are found in this habitat, but it is a favorite with nesting waterfowl as well as with rails, moorhens, and swamp sparrows. Drier areas support meadowlarks, bobolinks, and dickcissels.

- **Cane** up to thirteen feet high, which borders and sometimes invades cattail and grassy areas when standing water drains off for long periods. Cane seldom, if ever, gives way to other vegetation unless the water level changes. Few stands exist today, since the land on which they once grew was relatively easy to drain and cultivate. Of all the marsh habitats, cane supports the least wildlife, though it may be used as temporary shelter.

• **Earthen dikes,** the only true dry soil habitat in the marshes. Newly made dikes are first covered with dense stands of smartweed and velvetleaf, then with coarse weeds: burdock, cocklebur, milkweed, wild carrot, and thistles. If the dike remains undisturbed, the weeds are followed by shrubs, then by willow, cottonwood, and box elder trees. Keeping dikes clear of woody plants allows them to be used as roadways and encourages the low growth that shelters small mammals like voles and shrews.

Dikes and border thickets offer year-round homes for mammals, from the tiny shrew to the white-tailed deer, as well as for frogs, snakes, turtles, and the five-lined skink. Most of the larger mammals found in the marsh complex live on the dikes—or rather, in them: foxes build dens in the sloping sides and industrious woodchucks dig so many holes that some are taken over by raccoons, opossums, or rabbits. Waterline muskrat dens undermine the dikes. White-tailed deer, like people, use the dikes for travel.

• **Dense thickets** of saplings and shrubs such as buttonbush, wild currant, rough-leaved and alternate-leaved dogwood, hawthorn, elderberry, wild grape, prickly ash, staghorn sumac, and wild rose. Except for buttonbush, which likes its feet wet, these grow in any open space where the soil is fairly dry, on dikes or beside forests, roads, or ditches. The many seed- and berry-bearing plants make thickets convenient places for songbirds to nest, and a great many species build in them. Along the edges, tall herbaceous plants brighten the marsh with color: blue vervain, goldenrod, wild sunflowers, asters, pokeberry, and swamp thistle.

• **Bordering swamp forests,** now represented by comparatively few woodlots. The soil of a swamp forest is saturated but not completely covered with water except in spring and summer. Trees that can grow under these conditions include maples, locust, wild cherry, hawthorn, ash, hickory, sycamore, basswood, hackberry, a few black walnuts, and five kinds of oak. American elms, once common, have been eliminated by Dutch elm disease. Swamp forests house a few species of birds, such as orioles, woodpeckers, and nesting hawks and owls, but very few flowering plants grow there. The most common mammals are rabbits, raccoons, opossums, and white-footed mice.

• **Sand ridges**, the original natural outer barriers for all of the marshes. Before stone dikes were installed at the water's edge, the wet sand above the high water line held a variety of beach plants. These plants are now gone, although some of the others in this niche still grow at the borders of the marshes, among them sedges, flowering rush, evening primrose, bur marigold, and goose tansy, or silverweed.

Why should all this abundance of life be preserved? What good is it to us? Apart from their intrinsic aesthetic value, illustrated in the accompanying photographs, primitive wetlands may have some surprisingly practical uses. In a masterful—and prophetic—1960 essay, botanist Thomas Morley of the University of Minnesota enumerated some of these uses. Wetlands are havens for rare or nearly extinct species, including bacteria and fungi that may become sources of antibiotics and other drugs or means of natural pest control. Anticipating the stampede to the rainforests by today's pharmaceutical companies, Morley added, "No one can say what improbable organism in what unlikely part of the globe will eventually prove to have needed qualities."

Wetlands are among the few remaining storehouses for breeding and seed stock, preserving genetic diversity in a world where hundreds of species are daily becoming extinct, even as our pool of domesticated plants and animals shrinks to a few commercially viable varieties. Confronted with insects and bacteria increasingly resistant to control, confronted with evidence suggesting that our climate may be changing, plant and animal breeders are turning to "wild types" in search of strains resistant to disease, pests, and drought. Wetlands, which foster vigorous competition for survival, encourage the evolution of hardy, well-adapted species. Morley expresses in a single sentence the folly of blindly rearranging the face of the earth: "*We may endanger our future welfare by destroying what we think at the time to be unnecessary.*"

If the marshes and other wetlands are so indispensable to the whole of life, from microorganism to mankind, why are hunting and trapping permitted even in federal and state marshes? Because controlled hunting and trapping are not contrary to good conservation practices. No hunter or trapper in the past seventy-five years has brought about a serious drain on any form of marsh wildlife. In fact, amateur and professional reptile and amphibian collectors are a greater threat to the survival of these creatures than hunters ever were to any species.

Not only have hunters done no harm to marsh wildlife, but they have played a vital role in preserving the marshlands. Their license fees, duck stamps, and excise taxes have enabled federal and state agencies to purchase and maintain areas favorable to wildlife, saving them from sure destruction. Moreover, these days it is not only hunters and trappers who enjoy the marshes. According to the U.S. Fish and Wildlife Service, the most popular activity in marsh areas nationwide is sightseeing—which must include birdwatching—followed by fishing, picnicking, camping, and hunting. Locally, the Ottawa National Wildlife Refuge and the state-owned Magee Marsh are well-known to birders, nature photographers, hikers, and students pursuing degrees in wildlife biology. All this is a far cry from the days when all the Lake Erie marshes were privately owned and the marsh owners, wary of poachers, forbade anyone to enter their property.

Before the 1950s, when the state of Ohio acquired the Magee Marsh, I had access to only two marshes—Cedar Point Marsh and Michigan's Erie Marsh—and I was unusually fortunate. Most people not lucky enough to belong to one of the private shooting clubs could only park along the roads and hope to pick out a few samples of marsh life through their binoculars. Today, although its primary mission is protection of wildlife, the Ottawa Refuge offers nine and a half miles of trails winding through forest, wet prairie, and dike tops over more than six hundred acres. At any season of the year—except, of course, hunting season—a hiker can visit every type of habitat associated with a marsh, observing its life at close quarters. Ohio's state-owned Magee Marsh, which aims to accommodate both wildlife and people, features an observation platform, nature trails, and a public fishing area. But Magee's most popular attraction is the quarter-mile bird trail in a patch of woods that is a magnet for migrating songbirds. In spring, more birds visit the marshes than anywhere else in Ohio, and the raised wooden bird trail gives everyone, including the elderly and the handicapped, the opportunity to enjoy them.

The Nature Conservancy has estimated that a thousand acres of wetlands are lost every day. In this area, we are fortunate that both the state and federal governments are committed to preserving the marshes. As long as these lands are publicly owned and adequately protected with strong dikes, the marshes will be here for future generations to enjoy. Why preserve the marshes? For our continued well-being on this planet, for much-needed recreational areas, for our aesthetic enrichment, for the benefit of those who come after us—in short, to hold on to the irreplaceable.

PART I

A Wetlands Almanac

As season follows season through the years, the marsh follows an age-old pattern. Migrating birds arrive and depart; plants flourish, bloom, and die; mammals see their offspring develop into adults and go their way. A naturalist can almost foretell what will be happening on any given day in spring, summer, fall, or winter. Of course, each visit brings its own encounters: a strange bird, an unusually striking flower, an intimate glimpse into the family life of a common but secretive animal—delightful uncertainty within a broad pattern of predictability.

I have put together this almanac to give the reader a sense of what goes on in the marsh at each season. We all can respond to a stirring melody, but increasing our knowledge of music greatly enhances our enjoyment; just so, the more we know about the hidden life of the marsh, the more we can appreciate what is before our eyes—can *see* what we are looking at. Most of these observations would be true of any marsh near the Great Lakes, but they are based on my experiences in the Lake Erie marshes between Toledo and Port Clinton, Ohio. Some readers may prefer to begin with the histories of these marshes, which are told in Part II.

Today's visitors will not find the marshes exactly the same as I have seen them during past decades. During the fierce storms of 1973 and 1974, waves smashed dikes and rolled over beaches, destroying flora and fauna, suddenly turning wetlands into bays or ponds. The average level of Lake Erie remained high throughout the 1980s, and most of the largest marshes would have disappeared altogether if the old dikes had not been replaced by huge blocks of stone. Indeed, in two undiked sections of the shoreline, former marshlands are now alternately either bays or mudflats, depending on the lake level (see p. 116). By 1990, however, the diked wetlands were returning to their former condition.

January 1 to February 15

Shorthand on the Snow

In winter, the marsh itself is barren and desolate. Cattails, reeds, and long grasses are a brown, whispering sea tossed by an icy wind. In open places, drifting snow piles up in ridges against the cattails. There are few signs of life, since food is extremely scarce. Only an occasional mound of vegetation, like a miniature igloo, indicates that muskrats are still here. Sometimes a set of tracks can be seen in the snow, where a hungry muskrat left its shelter in search of food. The staggered prints are distinctive: wide-spread toes border the reversed curved line left by a dragging tail. The only other indications that last summer's cattail village was a thriving community are the oval, bulky nests of marsh wrens and the neater structures of red-winged blackbirds, now capped with snow.*

But the brushy borders of the marsh, the weed-grown dikes, and the occasional patches of neighboring swamp forest provide living quarters for many kinds of wildlife. Sandwiched between the frozen marsh and the bare, open fields of the farms crowding the wetlands, they form a sanctuary for hard-pressed birds and mammals. Here food is plentiful: seeds, nuts, acorns, and wild fruits such as grape and sumac. Numerous shrubs, together with the arched tangles of last summer's tall weeds, provide birds and mammals with shelter from the cold and protection from their enemies.

It is after a snowfall that the number of furred and feathered

---

*Throughout the book, I have used common names for all birds—"crow," for example, rather than "American crow." When a common name has changed, the current name is followed by the former name in parentheses—for example, "tundra (whistling) swan." Appendix B (p. 193) is a complete list of area birds, with their official names and the frequency of their occurrence in each season.

residents is most evident. Their movements, and to some extent their identities, are written plainly on the snow. At such times the hiker, exploring terrain unmarked by human footprints, sees signs of life everywhere. Around the protruding stalks of seed-bearing weeds is a delicate tracery of tiny prints left by dark-eyed juncos, song sparrows, and tree sparrows. Larger markings indicate where they were joined by a few wintering red-winged blackbirds. Separating and converging trails show the passage of a flock of ring-necked pheasants, now all too rare. A growth of giant ragweed has become home for a gathering of cardinals; as they fly out, the bright red of the males contrasts pleasingly with the white background. Where evening primrose and wild sunflowers offer their stores of seeds, goldfinches, pine siskins, and—if the observer is fortunate—redpolls from the far north whisper together as they feed.

The most common mammal tracks are the two-dot, two-dash of the cottontail rabbit, progressing through the courses of its dinner: first, seeds from locust pods; next, kernels of spiny cockleburs; finally, to top off the meal, bark gnawed from willow saplings. Where the underlying grass is thick, the snow is covered with tiny tracks of voles, or short-tailed meadow mice, sometimes ending in a tunnel.

I once came upon a vole that had evidently been feeding on fermented hawthorn berries. It was trudging round and round, marking out a three-foot circle in the snow, so intent that it didn't look up as I approached. When I picked it up, it squirmed and squeaked but, contrary to a vole's normally feisty nature, made no move to bite my fingers. Back on the ground, it was quiet for a moment, then found its former track and resumed the same slow, dogged circling. I was familiar with the behavior of cedar waxwings after eating fermented mountain ash berries, but this was my first encounter with a drunken mammal.

In wooded areas the vole's place is taken by the gentle white-footed mouse, sometimes confused with the deer mouse. It is a good climber, often making its winter home in a well-preserved bird's nest, which it covers over with grasses. If a watcher is cautious and moves slowly, he can disturb this velvet-eyed, long-tailed mouse just enough to coax it from its warm lair. Sometimes it will move only a few inches before clinging to a nearby twig, and will allow its tail to be touched. For some reason difficult to explain, a man feels very close to nature when a small wild creature curls its tail around his finger.

The path of a night-prowling raccoon is easily tracked by the marks of its distinctive hind paws, which have been compared to the feet of a human baby. But the opossum's trail is more difficult to decipher, its footprints irregular and haphazard as if it were returning from an all-night party. In contrast, skunk trails appear to be made soberly, carefully and deliberately, showing every claw, as one would expect from an animal with all the assurance in the world. Squirrel tracks are common in the woodlots, showing plainly where a fox squirrel has left its den in a hollow tree, crossed the snow, and dug out a nut or acorn buried weeks before. And weaving everywhere, in and out of the thickets, through the woods and over paths on top of the dikes, are the evenly spaced, precision-made tracks of the red fox, one foot set methodically ahead of the other.

A special discovery is a series of opposite markings imprinted at comparatively wide-spaced intervals. These could have been made only by a mink. One can almost picture it leaping sinuously over the snow as it leaves its home under a stump on the canal bank, looking for a mouse, a wandering muskrat, or an unwary short-tailed shrew.

Although the mink is pictured as a most efficient predator, ounce for ounce and inch for inch the short-tailed shrew and its smaller relative, the old field shrew, make the mink seem as harmless as a rabbit. The larger shrew (*Blarina b. brevicauda*) weighs an ounce and measures six inches plus a one-inch tail. The smaller shrew (*Cryptotis parva*) is three and a half inches and half an ounce of cold fury. Both shrews are mouse gray, with small black eyes nearly hidden by fur. Their eyesight is poor and they depend largely on a keen sense of smell to locate food. As a rule, neither species lives longer than two years.

The shrew's mission on earth is to eat: its metabolism is so rapid that it must consume its own weight in food every three or four hours. When all other sources of food fail, it turns to cannibalism.

Fortunately, shrews are omnivorous, although they prefer live prey when they can catch it. They eat berries, nuts, grain, insects, earthworms, snails, salamanders, snakes, probably frogs, all types of mice and, no doubt, chipmunks. In addition to a formidable set of teeth and a ferocious disposition, the shrew's saliva contains a poison that slows a victim's heart and lung action, making it the only poisonous mammal in North America. The vole greatly outweighs the old field shrew, but hasn't a chance of surviving when it is caught in the open. Shrews are protected to some extent against predatory birds and mammals by a gland in each hind leg that releases a strong odor of musk, discouraging attacks. This may account for the dead but uneaten shrews often found on top of dikes, apparently dropped by hawks or owls.

On one occasion, when I was moving slowly and quietly along the border of a marsh, I heard a confused chorus of high-pitched squeaks. Peering between the tall marsh grasses, I saw two small shrews fighting. They rolled over and over in the soft snow, each attempting to get a firm hold on the other. Occasionally one would break loose and try to escape, but each time the other was on it in a flash.

Apparently, shrews are not harmed by their own venom, because neither appeared to let up. Finally, one maneuvered around until it grasped the other by the throat with its strong jaws. There was a spot of crimson on the white snow; the loser quivered a few times, then died. Wasting no time, the victor began to drag its meal into a burrow. I found myself marveling at the savagery of the encounter, and grateful that shrews don't grow to the size of grizzly bears.

With so many mammals and birds about, numerous feathered predators can be expected. In winter, the number of hawks and owls seen over and beside the marshes reaches its yearly peak. Red-tailed hawks and northern harriers (marsh hawks) are joined by spectacular rough-legged hawks from the far north. The graceful kestrels skirting the edges of the marsh are sometimes accompanied by northern or loggerhead shrikes, whose black and white colors match the landscape. Sometimes a Cooper's hawk on its southward journey decides that the food supply here is too good to leave behind.

Owls take over at night, after the hawks stop hunting. While the great horned owl searches the countryside for the larger mammals, the long-eared owl haunts the wood lots, and the short-eared owl sweeps over open grassy areas, sometimes even in daylight. The erratic flight of the short-ear is reminiscent of a rower alternately dipping his oars—a good field mark. At times this owl will go into a short dive and strike its long wings together below its breast up to ten times a second, creating a fluttering sound. This happens most frequently during courtship, but it may occur at almost any season. Screech owls, either red or gray, are excellent hunters, seeming to prefer small birds to mice. In some years, two widely contrasting varieties of owls move in from their northern homes: the tiny saw-whet and the spectacular snowy owl. Because they are not well-acquainted with humans, they show little fear.

Most of these predators will never be noticed by the hiker plodding through the snow. But occasionally, a set of footprints will end in a splash of blood and the marks of widespread pinions on the snow. Brutal? Perhaps. But food for predators does not come labeled and plastic-wrapped in supermarkets. If the average shopper had to kill and dress the meat that comes so neatly packaged, we might see a tremendous increase in vegetarianism. There is too much sentimentality informing our concept of predators.

Actually, predation is an inescapable part of life. The dragonfly eats the mosquito, the kingbird eats the dragonfly, the screech owl eats the kingbird, and the great horned owl eats the screech owl. Yet most predators feed on a limited number of animal species. Only humans indiscriminately destroy everything from ants to zebras.

February 16 to March 15

The Marsh Awakens

Long before the snow melts, signs of spring appear in the marshes. Willow branches turn bright gold, buds on maples and poplars swell to the bursting point, and a few pussy willows crawl from their brown sheaths. Doves, cardinals, and song sparrows test their voices after months of silence, joining the chorus of tree sparrows and juncos. The sun takes advantage of longer daylight hours to shrink the snow drifts and soften the ice covering the marsh pools. Often in late February open water appears at springs, at the mouths of streams, and at the ends of drain tiles. Spring-like weather may arrive any time between February 22 and March 10, triggered by the first warm southwest wind. Then nature moves swiftly over a large front, bringing hundreds of birds of many species migrating from the tropics to their spacious northern breeding grounds. The Lake Erie marshes are a convenient stopover on the way.

Waterfowl dominate this first migration. Mallards, black ducks, and streamlined pintails circle the marshes and nearby fields, looking for open water. Sonorous honks overhead reveal the first flocks of Canada geese. Most impressive is the vanguard of tundra (whistling) swans, whose beauty, grace, and dignity are seldom matched on the water or in the air. Unlike that of most water birds, the swan's call is pleasantly melodious. Swans are very talkative, and their voices often reveal their presence even though they may be hidden in the depths of the marsh.

Waterfowl soon reach their maximum numbers of the year. Diving ducks like the redhead, canvasback, lesser scaup, ring-neck, bufflehead, a few ruddies, and the three mergansers seek out the first open waters. In flooded fields bordering the marshes, American wigeon, gadwalls, green-winged teal, a few blue-winged teal, shovelers, and wood ducks feed on waste grain and weed seeds. All are dressed in their most brilliant plumage. An occasional cold front moves in, freezing the ponds and the puddles in the fields. But even though winter has apparently returned, the birds don't return south; they seem to sense that warm weather will be back in a few days.

A major reason for this great concentration of waterfowl is that two of the major North American flyways pass over the western end of Lake Erie: the Atlantic Flyway funnels birds from the Atlantic Coast to the western prairies, and the Mississippi Flyway follows the Mississippi Valley from the Gulf of Mexico to the western prairies and the Hudson Bay area.

Another spectacular migration buoyed by the southwest winds is that of the crows and hawks. Crows, which formerly moved across the marshlands in great numbers, are now fewer though still common. Hawk flights are limited to clear, warm days, when spiraling groups of red-tailed and red-shouldered hawks appear, usually accompanied by a few kestrels, Cooper's and rough-legged hawks, and harriers. These birds are reluctant to cross Lake Erie, since they depend on updrafts to keep them airborne with a minimum of effort, and cold or frozen waters don't give rise to these strong air currents. So the hawks pile up against the shore, then swing southwestward to cross the mouth of the Maumee River; from there they either continue into Michigan or cross the Detroit River into Canada.

When winter's hold is finally broken, the thickets, swamp forests, and fields bordering the marshes suddenly come alive with songbirds. Great gatherings of starlings, red-winged blackbirds and grackles—over a million at their peak—drift across the landscape like clouds of black smoke. Flocks of robins, joined now and then by a vivid bluebird, search the thickets and dike banks for food. The number of mourning doves and song sparrows is swelled by newcomers from the south. Toward the end of this period, brown-headed cowbirds and rusty blackbirds infiltrate the great throngs of red-wings.

In the fields, the first meadowlarks pipe, and, rarely, Lapland longspurs in their new mating plumage scour the cultivated lands. Groups of snow buntings, their wings flashing white, hurry past as though late for an appointment. Thousands of herring and ring-billed gulls, which left when the waters were frozen, return to feast on winter-killed fish. Birders search these flocks for rarer varieties like the glaucous, Iceland, and great black-backed gulls. Two shore-birds arrive: the killdeer, with the incessant shrill calls that have earned it the name *Charadrius vociferus vociferus,* and the woodcock, which drops into the boggy thickets like a shadow and is seldom seen again except during its spectacular mating flight.

In contrast to the well-publicized advent of most birds, a few arrive in silence. Increasing numbers of saw-whet, long-eared and short-eared owls stop off on their northward journey, but they are only seen when some enterprising birder spies one in a grapevine tangle or flushes a short-ear from a grassy portion of the marsh.

Even though the snow is newly melted and most of the birds are just part way on the journey to their summer homes, some are already nesting beside the marshes. A gigantic nest in a huge cottonwood houses the magnificent bald eagle, incubating its eggs. Red-tailed hawks and great horned owls are keeping eggs warm in the bordering woodlots.

In a barren, windswept field adjoining the marsh, a prairie horned lark has laid her eggs in a crude nest built in a shallow depression in the earth. Spring storms regularly blanket this courageous bird with snow, but she does not give up; she maintains life-preserving heat in her eggs through the whole of a very trying period. As if encouraging his mate to persevere, the male often circles in the air above, singing his flight song. Floating down from a somber, cloud-ridden sky, his tinkling, brittle notes blend fittingly with the drifting snowflakes.

March 16 to 31

When Wild Winds Blow

March is a month of sudden changes from warm sunshine to snow, as high and low pressure areas chase each other across the continent. Northwest gales sweep ominous mountains of clouds across the marshes. Sometimes the clouds seem almost to brush the treetops, and circling birds are hidden behind the masses. It is a great season for the outdoorsman who loves to thrust himself against a hurrying wind doing its best to snatch his hat. As he notes the two families of birds that best typify primitive America—waterfowl and hawks—he feels close to a world untouched by civilization.

By the end of the month, waterways are open and all the species of swans, geese, and ducks that regularly visit western Lake Erie have arrived. A hiker following dikes and marsh edges flushes ducks every few yards. In the open pools there may be as many as twenty species of waterfowl. Both blue and white snow geese are less common in spring than in autumn, but a few usually stop off. Waterfowl are great gossips on their northward flight, and the marsh is loud with their varied voices.

Some of the ducks—the hooded merganser, pintail, wigeon, wood duck, and the two forms of teal—are surprisingly colorful. Since these ducks are much tamer in spring, the wildlife photographer is alert for the best picture of the year. Now is the time to search for rarer species: barnacle goose, white-fronted goose, Eurasian wigeon, old-squaw duck, and the three scoters.

The departures of tundra swans and Canada geese are as impressive as their arrivals. I once witnessed this ceremony when about two thousand swans resumed their journey to the northwest. The day was cool, but quiet and sunny. At first the birds were scattered throughout the marsh, all seeming to call at once, with flocks drifting back and forth. But soon I discovered that they were gradually gathering on one large pool. Then, in the middle of the afternoon, with a great clamor and splash, they took to the air, circling and rising in a vast, swirling cloud. As if on signal, a group of a hundred or so split off and headed northward along the shoreline, followed at nearly regular intervals by other groups, until only a comparative few remained behind. As they faded into the distance, each group flying abreast, they became a series of parallel white lines.

Waterfowl are now joined by many other water birds: great blue herons and their white counterparts, the great egrets, black-crowned night herons, pied-billed and horned grebes, and those clowns of the marshes, the coots. A double-crested cormorant may drop by on occasion.

Overhead, on days when updrafts make flying easier, circling bands of red-tailed and red-shouldered hawks drift by. Turkey vultures, masters of aeronautics, cover a broad path as they glide effortlessly on an apparently aimless journey. Hawks that prey on birds are now more evident, and an observer may be thrilled to see a rare peregrine falcon or a merlin, or to watch a more common Cooper's or sharp-shinned hawk making a kill.

Along the trails, newly arrived songbirds join those already here. The rufous-sided towhee flashes its white tail feathers, a dignified fox sparrow quietly stands inspection, a phoebe darts through a cloud of gnats. Northern flickers announce their arrival with loud cries and rattling tattoos on suitable-sounding branches. Belted kingfishers hover over the canals, searching for an unwary small carp. And, all too rarely these days, a loggerhead shrike may appear.

Toward the end of the month, new voices compete with bird songs: the melodious notes of the chorus frogs, the bass grunting of leopard frogs, and occasionally the birdlike calls of spring peepers. On unusually warm days, a garter snake or a gentle DeKays snake may be found sunning itself, but normally water and fox snakes and turtles sleep to a later date. A few mourning cloak and red admiral butterflies are on the wing, looking very much out of place in a barren landscape, occasionally joined by a migration of those skilled predators, the dragonflies. Small mammals are active, especially at night. For several of them this is the mating season, and they lose much of their caution. The toll of rabbits, raccoons, opossums, and skunks on neighboring highways is disheartening.

April 1 to 15

Goodbye to Winter

Bands of very famous bird travelers arrive in April. These are the shorebirds, some of which have spent the winter on the pampas of Argentina and are now on their way to the tundra of the Arctic Circle, where they will raise their families. In the first group are greater and lesser yellowlegs, pectoral sandpipers, the common snipe, and, most famous of all, the golden plover.

The migration route of the golden plover has been worked out in great detail. Leaving Argentina, it moves northward across the Gulf of Mexico and up through North America. It mainly follows the Mississippi Valley, but on a path so broad that hundreds of birds pass over the western end of Lake Erie. The entire journey to the plover's breeding grounds on the Arctic tundra covers 8,000 miles. After the young are able to fly, they return to Argentina by the same inland route, while their parents travel eastward to Labrador. Here they take off on a 1,400-mile overseas flight to South

America. The annual round-trip mileage flown by adults is 18,000 miles—all very wasteful by human standards.

While these travelers are crossing continents, another shorebird, the woodcock, has already settled down to business: it has laid its clutch of eggs in the swamp forests and boggy thickets bordering the Lake Erie marshes. Killdeer are also nesting, in open fields and sometimes even in suburban driveways. Intruders are met with the incessant screaming of the male as he circles overhead, while the female may draw attention from her brood by fluttering some distance away, faking a broken wing.

At this season the rapid succession of arrivals and departures is reminiscent of a major airport. The great flocks of waterfowl in the marsh dwindle as many head northward, and the number of coots reaches its peak. About the middle of the month, the moorhen (gallinule) appears, and the sora announces its arrival with a whinnying call from its hideout in the vegetation. An American bittern standing upright beside the water's edge freezes as a hiker passes by, convinced that its camouflage pattern renders it invisible. Herring and ring-billed gulls increase in numbers, and soon the smaller, black-headed Bonaparte's gulls join them. Coursing over the pools, purple martins and tree swallows feed on gnats.

In the swamp forests and shrubby edges of the wetlands, migrant songbirds are arriving daily. Some are difficult to find: the secretive yellow-bellied sapsucker, the winter wren creeping mouse-like through the grapevine tangles, and the hermit thrush, quiet as a shadow along the trails. Others, like the brown thrasher and the white-throated, vesper, field, and swamp sparrows, make no effort to conceal themselves. Toward the end of this period, a great wave of golden-crowned and ruby-crowned kinglets moves in, usually accompanied by brown creepers. These diminutive birds are remarkably fearless and can be approached almost to arm's length.

Increasing warmth from the sun coaxes out the earliest wildflowers: first the tiny mouse-eared chickweed, then the cut-leaved and purple spring cress. Marsh grasses and new blades of cattail form a fringe of green around the pools and beside the dikes. Pussy willows are golden and a few reddish catkins dangle from poplar twigs.

Muskrats are enjoying the bright sunshine and wander openly

along the shorelines, searching for food. Woodchucks have dug their way out of their winter dens; they and the raccoons love to lie along tree branches, apparently afflicted with a touch of spring fever.

Chorus and leopard frogs now sing continuously. Sometimes they are accompanied by the high-pitched treble of an American toad, but cricket frogs, green frogs, and bullfrogs remain dormant. The DeKays and garter snakes are now joined by water snakes, while other varieties sleep on. Turtles, too, have not yet ventured out of the mud and vegetation, perhaps because their shelters are the last to feel the sun's heat.

Nature seems to spend most of this period consolidating its gains, eliminating the last signs of winter and preparing for the miracles of spring.

April 16 to 30

Drake Meets Duck

Courting time has come to cattail village, which resembles one huge Lover's Lane. Not only are the summer residents of the wetlands pairing off, but also the ducks and Canada geese that will make their homes in the North and Northwest. No one knows why a drake singles out, from thousands of ducks, a particular female as the object of his affections—any more than we can explain the chemistry between a man and a woman. Waterfowl are not promiscuous, and certain formalities are in order before mating takes place. Some of these may appear ludicrous to us, but they are performed with utter gravity. (I resist the temptation to compare them to some human courting performances.)

Canada geese mate for life. A courting goose preserves his dignity, confining himself to strutting toward the female of his choice with head extended close to the ground and a lecherous look in his eye. He does get excited, however, when a competing gander approaches, beating the intruder with his wings and attempting to grasp the enemy's head in his beak.

The courtship of puddle ducks shows little variation among

species. Both sexes indulge in preliminary bobbing and bowing and caressing each other's feathers with their bills after which the female takes off on a wild flight circling the marsh. The male follows her, joined by as many as four other swains, filling the air with quacks and whistles. One by one the males drop out, until only the successful one remains—sort of an aerial elimination contest. Only one species is conspicuously polygamous: the female shoveler is not averse to sharing her affections with two males.

Drakes of the diving species are much quieter, adopting several stereotyped gestures common to nearly all in this family. They drop their heads and press their bills against their breasts, then throw their heads backward until the bill touches the rump feathers. Those that wear a well-defined crest raise and lower it like a fan; the others simply ruffle their scalp feathers. Sometimes diving becomes part of the courtship ritual.

The hooded merganser and the smaller divers are the most demonstrative. The bufflehead's performance is spectacular. He is a small, graceful bird, a talented diver able to disappear at a snap of the fingers. He is also very quarrelsome and will not tolerate other drakes near him when he is courting. First he swims near and around the females, bill pointing upward, neck extended, head puffed out to twice its normal size. Then, selecting one of the hens for individual attention, he rears up on the water, erect on his feet and tail, and struts toward her, his bill drawn down upon his swelling bosom. Suddenly he dives and, upon surfacing, moves nonchalantly toward another hen and repeats the performance. This continues until a female follows him and a pact is made.

The drake ruddy duck swims around a female as though conscious of his charm. He shows off by tilting his tail forward until it shades his back, stretching his neck to its fullest extent, then drawing it back close to his body. He puffs out his red chest by inflating a special air sac in his neck, and slaps his brilliant blue beak against it. If this formula is not successful, he presses his bill against the raised feathers of his back and, with tail submerged, scoots upright over the surface with surprising speed. Sometimes he kicks both feet at once, making the water spurt behind him. The female signifies that she is impressed by stretching out on the surface with head, neck, and bill at water level. Despite the drake ruddy duck's apparently absurd contortions, he deserves respect: he is the only drake that helps raise his offspring.

At least once in this life, everyone who loves the outdoors should witness the courtship of the American coot, commonly known along Lake Erie as the mudhen. This black waterbird with the large, chalk-white beak is consistently and comically awkward, whether swimming, diving, or walking, and does nothing during courtship to improve its image. After a few preliminaries, in which the male swims around the female, clucking softly, she swims away from him. When the distance between them is about fifteen feet, he begins to follow. With his head, neck, and beak lying flat on the water, his rump elevated and puffed out, he weaves from side to side. Apparently in order to appear larger, he elevates his wing tips over his back and spreads his tail widely, displaying to the utmost the white markings on each side.

The hen assumes a similar pose as he approaches. Then the chase begins. She splashes mightily with her large feet as though desperately trying to escape, but at the same time takes care not to move quickly enough to discourage him. When it appears he must overtake her, she raises her body and patters swiftly over the water, her suitor only a few feet behind. A loud chorus of "kuk, kuk, kuk" adds mudhen music to the commotion. When he reaches her, they both drop back in the water and begin calmly to preen each other's feathers.

While these courtships are taking place on the water, another is being acted out in the air above the grassy portion of the wetlands. A male harrier, or marsh hawk, is performing a series of aerial acrobatics to impress his intended mate, who flies just above the cattails or perches on a muskrat house. He first climbs to a height of several hundred feet, then dives headlong toward the earth, twisting, turning, sometimes somersaulting as though completely out of control. Close to the ground he pulls out of his dive, ascends, and repeats the performance, sometimes two dozen times in a row. All during the display, he screams constantly in a shrill voice. Unfortunately, this scene is becoming all too rare, as harriers continue their rapid decline of the past three decades.

Meanwhile, more feathered travelers pour in. Some, like the moorhen and the king and Virginia rails, being anything but taciturn, are heard before they are seen. The larger members of the heron family increase in numbers and are joined by the green-backed (little green) herons and least bitterns. Spotted and solitary sandpipers arrive, but they prefer to be alone and seldom join the flocks of other species. Common and Caspian terns, looking sleek and wellgroomed with black caps and blood-red beaks, patrol the ponds looking for minnows. Another fish-eater, the handsome osprey, occasionally hangs suspended in the air on rapidly beating pinions as it prepares to dive for a goldfish or a carp.

A spectacular event of this period is the northward push of broad-winged hawks. Like the migrating hawks of early spring, they do not like to cross the cold water of Lake Erie, and they turn southward to circle the west end of the lake. With them is a sprinkling of immature red-tailed, red-shouldered, rough-legged, sharp-shinned, and Cooper's hawks. These flights can involve most impressive numbers; for instance, on the weekend of April 26 and 27, 1969, 5,000 hawks were counted at one location, of which 4,500 were broad-wings and 400 sharp-shins. Until they were protected by state and federal laws, hawks were very often slaughtered by gunners during these concentrations, in the mistaken belief that killing predators could increase the game supply.

Gnats and midges are hatching in the shallow pools, drawing flocks of swallows. Barn, rough-winged, bank, and occasionally cliff swallows join the tree swallows and martins. Feeding higher up are chimney swifts and, now and then, a bat. Toward the end of the month, two large flycatchers appear, the great crested flycatcher and the kingbird. A third, the whip-poor-will, may be flushed from the ground or a fallen log. Overnight, sputtering house wrens become plentiful and dainty blue-gray gnatcatchers join their cousins the kinglets. Cedar waxwings and throngs of lemon-colored, melodious goldfinches drift along as though in no great hurry. At the same time, great companies of boisterous, strident blue jays, sometimes as many as five hundred, sweep back and forth along the outer beaches, seemingly undecided about which route to take.

A querulous gray catbird complains from a thicket, its notes shrill in comparison with the pure, high-pitched whistles of a band of white-throated sparrows. The forest borders are loud with red-headed woodpeckers and flickers. But birdwatchers are most intrigued with the first of the warblers, those small, varicolored birds that will pass through the marsh borders by the thousands in May. In the vanguard are yellow-rumped (myrtle), yellow, black-and-white, and palm warblers, yellow-throats, and northern water-thrushes, accompanied by a few species of vireos.

Gradually the forest floor becomes carpeted with wildflowers: violets in purple, white, and yellow, quaint Dutchman's breeches (pronounced "britches"), a multitude of spring beauties, golden buttercups, and white and yellow adder's tongues. Red maples and box elder trees are in bloom, and buds of buckeye and hickory are bursting. Hundreds of red-brown tassles decorate the poplars, and miniature leaves have sprouted on willows and box elders.

Snakes and turtles of several species sun themselves on every available log, and green frogs and huge bullfrogs have finally stirred from their winter sleep in the boggy depths. Since they are still sluggish, this is a perilous time for them.

Before the marshes were enclosed by dikes, April was a period of fish migration, when many species moved in from Lake Erie to spawn. First would come the great northern pike, followed by yellow perch, crappies, largemouth black bass, bluegills, pumpkinseed sunfish, channel catfish, several forms of minnows, including the colorful mudminnow, and black, yellow, and brown bullheads.

In a fish beauty contest, bullheads would undoubtedly rank at the bottom of the list: not only are they ugly and without scales, but they come armed with three needle-sharp fins just behind the head. These fins carry a mildly poisonous substance that stings like wasp venom when they puncture the skin of a careless or inexperienced angler. But in spite of all these drawbacks, the bullhead is very popular for family fishing. Why? Because even the youngest angler can catch it.

Many a fisherman who prides himself on his impeccable technique and sophisticated tackle will admit that the first fish he ever caught was a bullhead. It can be taken from the bank of a river or canal, it will devour anything edible, and once it grabs the bait it never lets go—in fact, the hardest part of bullhead fishing can be the struggle to extract the hook. Almost any sort of tackle will do. In my early days we used "throw lines" coiled at our feet, with a heavy sinker tied to the end and two hooks above. We would swing the end in a circle a few times and hurl it out into the canal, usually with good results. We enjoyed eating our catches, too, because bullheads have a rich flavor and relatively few bones.

Today, bullhead fishing is still good on Ward's Canal and Turtle Creek. The best place for families is the mile-long road beside Metzger Marsh Bay. I stopped by the road recently to watch several youngsters in action, and was reminded of the times my three children and I fished together. On one occasion, my oldest hooked a large bullhead and seemed uncertain of what to do next. When we shouted, "Pull it in!" he turned around with the rod over his shoulder and ran up the bank as fast as his six-year-old legs would go; behind him, the fish burst out of the water and bounced up the bank as if it were chasing him.

A bit later on that trip, this same son hooked a large bullhead and—obviously remembering the result of his earlier dash—began to pull up his fishpole with agonizing patience and deliberation. Afraid he would lose it, I took his rod and hauled in the fish. Later I said to him, "That was a nice fish you caught." "*I* didn't catch it," he said. "*You* did." I never forgot that lesson: Sit on your hands and let your child land his own fish. If it's a bullhead, chances are he will.

Mammals of all sizes are busy with family chores, some already introducing their youngsters to an interesting and dangerous world. Small rabbits and groundhogs run ahead of a hiker on the path, but skunks and foxes keep their offspring in dens until they are fairly well-grown, and raccoons most often guard their young in hollow trees far above the earth. Voles and shrews build their nests of grass beneath planks or logs. Almost every discarded sign I ever saw, wood or metal, protected its quota of nests. White-footed mice also like shelter, but I have found their nests in the open as well.

All must be especially wary, keeping little ones under cover.

Kestrels, harriers, and red-tailed hawks, great horned and screech owls—all have fledglings to feed. There are few, if any, quiet hours in cattail village. Action is at fever pitch day and night.

May 1 to 15

There's Music in the Air

This is the most beautiful time of year in the marsh. Butterflies and dragonflies are on the wing, and songbirds and even birds of prey wear their most brilliant plumage. The colors of some of the waterfowl rival even the multi-hued warblers. Many trees and shrubs are covered with blossoms, and hawthorn, plum, cherry, and wild crab fill the air with perfume. Tiny green leaves have crept out of their buds, giving a misty look to the groves. The ground beneath the wooded patches is splendid with wildflowers, each one small, but collectively a marvelous, once-a-year color pageant. Marshes are fringed with the delicate emerald of new grasses, sedges, and cattails. Some dikes are carpeted with golden dandelions.

So engrossed is the eye in capturing every facet of the scene that the ear takes second place. Then, almost suddenly, it becomes attuned to a symphony of wild sounds: the piping of kinglets and warblers; the flute-like notes of doves, cardinals, titmice, and thrushes; the baritone of flickers, grackles, and rusty blackbirds. The brown thrasher is soloist, carrying the theme. It repeats each phrase two or three times, like a songwriter at a piano. Supporting with contrapuntal melodies is a choir of brightly-costumed birds: flaming orioles, scarlet tanagers, rosebreasted grosbeaks, indigo buntings, and goldfinches. Yet some of the plainest in their brown robes are among the best musicians: field, vesper, white-throated, white-crowned, and song sparrows. The snare drum of woodpeckers is not discordant, but blends well with the wild chorus. From bordering meadows, the distant sweet harmonies of bobolinks and eastern and western meadowlarks drift in with echo-like persistence.

Bird songs are so spontaneous and exultant that the sober scientific assessment that they sing only to attract a mate or to proclaim territorial breeding limits seems most inadequate. If such reasons explain each trill, why do lone males, hundreds of miles from their breeding grounds, sing constantly? Would it not be just as reasonable to suppose that they sing because "the winter is past" and a new splendid year is underway?

Cattail village also has concerts strongly reminiscent of Stravinsky. The dissonant music of herons, rails, and coots, of redwinged blackbirds and marsh wrens, reflects the primitive nature of the surroundings. An important section is made up of frogs—leopard, chorus, cricket, and green—dominated by the booming tympani of the bullfrog. And always blending with this unique medley are the minor-key murmur of the winds through the swamp forest and the rustling of last year's dried cattails and reeds. From the beaches comes the regular beat of waves crashing on the barriers in faultless precision, like a giant metronome.

The marshes and their wooded borders are among the finest places in Ohio to see birds. In spring, the Lake Erie shoreline from Toledo to Marblehead is rated the fifth largest songbird concentration area in the United States, and in the first two weeks of May alone some sixty species arrive from the south to join those already here. The reason for these large concentrations is Lake Erie, always a formidable barrier to migrating birds. If a cold or storm front moves in, unbelievable numbers of birds can accumulate along the shoreline. Sometimes birders can even see a reverse migration when flocks start to cross the lake, lose heart, and return to the mainland. This is often an excellent time to find rarer birds, either because they have been blown off course or, in the case of southern species, because they have traveled too far northward. However, a northeast wind keeps all birds, even hawks, away from the marshlands.

When I first began to study birds in the marshes, in 1927, Little Cedar Point was the major concentration area for migrating birds. The Cedar Point Marsh is triangular in shape, with a clump of trees and a sandbar at its northwest point (see map 5 on pp. 120–121). In those early days, a dense forest bordered the entire southern edge of the marsh, and the outer beach to the north was lined with trees; these two wooded areas sheltered northbound birds getting ready to cross the mouth of the Maumee River and head for the Michigan shoreline. Even large insects such as dragonflies, monarch butterflies, and bumblebees followed this route. Eventually, however, the trees leading to the Point were destroyed—those along the beach by water, those on the south by lumbering—and by 1965, there were no noticeable concentrations at Little Cedar Point.

For many years there was no easily accessible public bird trail in any concentration area. The nine-and-a-half-mile Blue Heron Trail in the Ottawa Refuge, which passes through swamp forest, wet prairie, and marshes, is an excellent place to see a great variety of birds, but it lacks concentrations. The Navarre Marsh, home of the Davis-Besse nuclear plant, attracts the area's greatest concentrations of migrant and breeding birds, but visitors are allowed only in groups and permission to enter must be arranged in advance. Fortunately, the bird trail in the state-owned Magee Marsh is laid out in a small patch of trees that for some reason attracts thousands of migrating birds representing dozens of species.

This is the time of year when throngs of warblers flood the bird trail. All but one, the yellow-breasted chat, are smaller than a house sparrow. Because they come in so many colors and patterns and are seldom, if ever, motionless, they present a special challenge to the birder. Thirty-eight species of warblers have been reported, but the number of variants to identify is much higher because of females and first-year males in different plumages. A bird identification manual is indispensable equipment for a May bird hike. Songs are also a great help, but it is a most unusual person who can remember all the warbler songs. A birder can make friends on the trail by pointing out a rare bird.

The fickle nature of spring weather means that one day the bird trail may contain literally hundreds of birds, the next day only a handful, most of them residents. Until the middle of May there will be either feast or famine, usually the latter. Birders take their chances, but when they do encounter a collision of weather fronts that drops clouds of birds along the beaches, they experience a day they will always remember.

May 16 to 31

Prelude to Summer

In many years, the weather during the first two weeks of May is more like early April: cold fronts, trailing each other, bring chill rains and even a trace of snow. These periods are occasionally so extended that purple martins and perhaps other swallows die of starvation and exposure. The last half of May, however, is normally warm and sunny. Warblers, vireos, flycatchers, some thrushes, and several other species remain in the South, waiting for favorable migrating conditions. When the time comes, they move northward in a great wave, flooding the marshes and their borders. If the weather remains to their liking, they do not linger but pass on quickly to more northerly places.

By now, except for shorebirds, all the expected water birds are here and many have already moved on. Late May migrants are virtually all songbirds, including a marked flight of nighthawks and black-billed and yellow-billed cuckoos, together with Connecticut, blackpoll, Wilson's, and mourning warblers. Rare birds that may appear are the dashing yellow-headed blackbird and the shy sharp-tailed and Le Conte's sparrows.

Bird identification is complicated by a wave of small flycatchers that resemble each other closely. Two may be distinguished by size and voice: the eastern wood pewee with its plaintive "pee-a-wee" wail, and the diminutive least flycatcher, which chants "chebec" emphatically and well-nigh continually. The yellow-bellied flycatcher can be identified in reasonable light by anyone who can tell yellow from gray. But the rare alder flycatcher is difficult to identify, and the look-alike Acadian and willow flycatchers present a problem even to the experienced observer.

During the breeding season the Acadian flycatcher lives in the swamp forests, the willow flycatcher in the thickets beside the marsh. During migration, however, they are seen together. A sure and convenient method of distinguishing them is by their staccato calls. The Acadian says "pee-e-yuk" with the last syllable heavily accented; the willow shouts "whis-kee," sometimes "wish-beer"—

at least to my ears. Rounding out the list of flycatchers is the olive-sided, normally perched high on a bare stub, announcing over and over, "free beer." These interpretations have an obviously alcoholic flavor, but such thoughts are easily prompted by a long, hot hike. And they certainly approximate the actual sounds—which is more than can be said for the unimaginative birder who rendered the hauntingly beautiful song of the white-throated sparrow as "Old Sam Peabody, Peabody, Peabody."

A resident warbler of the marshes, much sought after by birders, is the striking golden prothonotary. No doubt in earlier times it was far more numerous because its habitat was more extensive. Today, as far as we know, its population has dwindled to one or two pairs in each large marsh, though this may be more pessimistic than the facts warrant, since few bird surveys have been made in the privately-owned marshes east of the Toussaint River. In the last decade, the prothonotary has appeared most often in the Navarre Marsh and on both sides of the mouth of Crane Creek. Occasionally, singing males are seen near the Stange Road Bridge over the creek in the Ottawa Refuge and near Little Cedar Point. I first found prothonotary warblers in the Magee Marsh along Lake Erie on June 2, 1928. On that occasion, a pair was using the abandoned nest of a downy woodpecker, drilled in a water-killed willow.

The most unusual prothonotary nest I ever saw was near the mouth of Crane Creek on June 16, 1930. Kenneth Byers, whose family maintained a commercial fishing base on the west bank of the stream, showed me the nest in one of the net sheds. It was built in a paper sack half-filled with staples, resting on a shelf near a window, and it contained four fledglings. Ken also produced a coffee can, a cheese box, and a small lard pail that had been used in previous years. That same day in the adjoining Magee Marsh we found two normal nests built in holes drilled by downy woodpeckers.

In recent years, various helpful people have set out nesting boxes in an attempt to attract more of these colorful warblers. Although in the past prothonotaries have occasionally used boxes or similar objects (years ago a pair nested in a mailbox), they continue to prefer cavities in trees. The single recent exception, according to biologist Mark Shieldcastle, is their use of boxes set up in patches of buttonbush in Navarre Marsh. These boxes were used even though the area is bordered by suitable trees.

In contrast, tree swallows, which breed commonly beside the marshes, use all types of containers, even wood duck breeding boxes. No doubt they compete with prothonotaries for tree cavities. Recently, Magee Marsh managers installed a series of boxes on both sides of the paved road leading through the marsh to the Lake Erie shore. In their first season, virtually all were occupied by tree swallows.

As May draws to a close, shorebird migration through and alongside the marshes accelerates. Different varieties now make up the flocks, replacing earlier migrants that have moved on. Many wear showy costumes, especially short-billed and long-billed dowitchers and red knots, with their red-brown breasts, dunlins, the bizarre ruddy turnstone, and the dignified and spectacular black-bellied plover.

It is a time when rarer shorebirds pass through: Wilson's and red-necked phalaropes; white-rumped, Baird's, and western sandpipers; and the larger species such as whimbrels, willets, marbled and Hudsonian godwits, and, occasionally, avocets. Most exciting of all is the very rare European ruff. Birders traditionally set aside the Memorial Day weekend as the time to make a spring survey of shorebirds.

Sometimes—and these are red-letter days—southern herons reach the peak of their migration path and end up in a Lake Erie marsh: the little blue heron in adult plumage, the yellow-crowned night heron (which rarely stays to breed), the startling glossy and white-faced ibis, the snowy egret, and, rarest of all, the tricolored (Louisiana) heron. The cattle egret was first seen locally in 1960. This bird introduced itself to the Americas from the African plains, where it feeds on insects, often climbing on the backs of large mammals in search of ticks. At some unknown time in the late 1800s, cattle egrets crossed the Atlantic Ocean and gained a foothold in South America, gradually infiltrating northward until they are now found in every state except Alaska.

The more showy insects are now abundant. Fragile damselflies and a host of voracious dragonflies in several sizes and colors patrol the ponds. Most of the showy butterflies are also on the wing, but mosquitoes are not yet numerous enough to be a nuisance except in the swamp forests.

Stopping on a warm, sunny day to rest on a fallen log, a hiker may hear a faint rustling and look down to find a five-lined skink sharing the improvised bench. The only lizard native to northwestern Ohio, the skink is found most often along the Lake Erie beaches. The juvenile skink is especially attractive, sporting a bright blue tail—which, incidentally, breaks off if you catch hold of it. Skinks are gentle, harmless, and a most interesting part of our marsh fauna. They should be strictly protected from collectors and those who want them for pets.

The earliest of the forest wildflowers are now brown and spent. But in full bloom at the edge of the deep woods and in the thickets are two kinds of white trillium, wild phlox, wild geranium, spikenard, golden corydalis, and, on poorly drained ground, wild iris. Showy clumps of wild columbine not only delight the eye, but also attract ruby-throated hummingbirds. Watching these birds, we discover that the breathtakingly beautiful male is impolite, even cruel, to his intended mate, driving her away from the blossoms. Hummingbirds are also frequent visitors to the fragrant hawthorn, buckeye, redbud, and wild cherry blossoms. The marshes are still shaded with the brown of last year's cattail and cane, but the green of new growth is fast enveloping them. Globes of spatterdock, first of the water lilies, add a welcome splash of gold.

This is a good time to explore the marshes by way of the many dikes. Vegetation has not reached its full growth, troublesome insects are few, and the weather is normally pleasant. Birders find peak numbers of species, occasionally as many as 125. But the outdoorsman must be prepared for the short, vicious thunderstorms that can sweep over the wetlands. Heralded by the muttering of distant thunder, mile-high masses of dark clouds move across the sky, eating up the blue. The flat terrain makes the formations appear mountainous. The wind velocity increases, accompanied by a few spatters of rain. Then from the distance comes the subdued murmur of an approaching downpour. Woe to the hiker who has neglected to bring raingear, because there is no shelter.

Suddenly, the storm arrives. Sharp flashes of lightning strike the eye almost like blows. Peals of thunder seem to echo back and forth between the cloud cliffs like barrels of sound tumbling down the steep sides of the cumulus piles. Few people are not awed, even frightened, by this violence. In the marsh, cattails bow their heads and whisper, like worshipers at a fearful religious rite. Then, as suddenly as they gathered, the black formations are gone and the welcome sun prevails again.

Night Comes to the Marsh

The old marsh guides frequently spoke of the uncanny stillness of the marsh after dark. Much of the year this is true, but at certain seasons cattail village bustles at night. Birds, mammals, amphibians, reptiles, and fishes contribute to a potpourri of night sounds: loud calls, muffled wingbeats, bodies rustling through grass and leafy barriers, and constant splashing. Nocturnal activity reaches its peak toward the end of May when the moon is full.

All of us who feel a special attraction to the marsh should plan a night trip at least once, on one of the convenient dikes. A companion or two will enhance the experience, if everyone is patient enough to be quiet and motionless, even for long periods. The best time for us to begin the long hike is when the last glow is fading in the west. Then ducks and other water birds appear to lose their wings, and only the ghostly great egret stands out against the black cattails.

A knowledge of bird calls is a necessity; otherwise their conversation merges into a jumble of sounds that mean nothing. Few insects sing in May, and the only ones we hear are the millions of gnats with their steady, almost imperceptible hum. Our first bird is a green-backed heron leaving its perch on a snag at the water's edge and flying ahead to another vantage point. This it does again and again, each time harshly voicing its protest at our presence.

The marsh is now in darkness except for the light of the rising moon. From overhead comes a loud squawk, answered by another close by; a small company of black-crowned night herons from the West Sister Island heronry is coming in to feed. We mimic their calls, and they approach so close that for a split second we see them outlined against the moon.

Frogs are tuning up in the ponds and along the edges of the bank. Ahead we can hear the insect-like click of cricket frogs—the smallest amphibians in the marsh—and the hoarse croaking of leopard frogs. The calls die out as we pass and then pick up again as if to speed us on our way. A green frog releases its sharp *"gunk,"* like the twang of a plucked guitar string. This sound is easy to imitate and we carry on a conversation with an interested individual, never knowing what we are saying, what great news we may be spreading among the frogs. Then we hear the booming "jug-o-rum" of the bullfrog, giant amphibian of the wetlands. The sound has a reverberating quality that makes its origin difficult to locate. The first marsh visitors may have pictured all sorts of monsters as its source before they discovered the truth. And what a letdown it must have been!

Ancient peoples, and no doubt some not so ancient, have pictured large marshes as the home of hobgoblins, demons, trolls, and other weird supernatural creatures that prowl only at night. As we move deeper into the marsh, we can understand why. Loud derisive laughter, screams, splashes, squawks, and noises that could hardly be ascribed to any earthly creature are magnified by the darkness. We find ourselves whispering to each other for no reason at all. For a moment we almost regret that we know these outlandish noises are no more than the mating calls of coots, moorhens, rails, and the pied-billed grebe, whose clamor is much too great for its size. A killdeer shrills its name over and over again as it circles the pond holes, sounding for all the world like a disembodied spirit proclaiming its woes.

Waterfowl are very active on moonlit nights, and we can hear them from the choice feeding places where they gather to eat and get acquainted. At intervals a female mallard tells her location to listening males with her rolling call, most unmusical for so beautiful a fowl. Above, a chorus of mellow whistles reveals a flock of

American wigeon as they maneuver for a landing. A foursome of blue-winged teal adds a soft-voiced "peep-peep-peep." Cartoonists whose wild ducks all say "quack quack" would be amazed at the variety of these vocalizations. As we follow the dike away from the open pools, we hear far in the distance the honking of a few pairs of Canada geese about to join others in the marsh.

Now the cattails have closed in almost to the edge of the dike, and there is less activity. It is so quiet we hear the rustle of a snake as it crosses our path and splashes into the water. We know it is either a fox, water, or garter snake, so we resist the impulse to break the spell with our flashlights. We can hear the painted and Blanding's turtles diving into the water from stumps and logs as we go by.

We begin to pick out songs of land birds. They are supposed to be asleep, but the moon-madness is upon them and at intervals one will pipe sweet notes that seem out of place in the surrounding blackness. A yellowthroat starts out boldly, *"look at me, look at me,"* then falters as though afraid someone actually might. A song sparrow trills from a patch of weeds, a yellow-breasted chat injects a few phrases from its staccato repertory. A marsh wren chatters from the security of the cattails, rousing a red-winged blackbird and his harem into raucous protest. A feeding nighthawk "zeeps" from the dark sky above.

All along our path we have been conscious of muskrats busily engaged in family duties. At times we see one in the path of the moon, parting the water into a giant V as it swims. Then we hear the commotion of a larger mammal in the path ahead. For the first time we use flashlights, and two ruby eyes gleam—an opossum. For a moment it remains quiet and we can see its pink feet with widespread toes; we can even pick out its pink nose. Then it shambles awkwardly away, apparently in no great panic, as befits a member of a species that has seen mammoths, mastodons, and sabre-toothed tigers pass into oblivion.

At another point we become aware of a mammal on the bank. We believe it is a raccoon on its nightly prowl for anything edible from crawfish to eggs—avian or reptile—but it is much too skillful in avoiding discovery, and we never get very close.

As the night progresses, owls begin to make themselves known. The high-pitched, tremulous musical call of the screech owl, nothing at all like a screech, comes from a wooded island. Whoever named this bird must have been tone deaf. Its quavering whistle is easy to imitate, and we coax the bird so close we do not miss a note. A series of harsh, high-pitched screams indicates a barn owl searching for shrews and mice over the grassy parts of the dike. This species has become the rarest of our owls. Then from a distance come the bass viol tones of the great horned owl, sounding its gruff "whoo, whoo." It is the most skillful of all winged predators. At this season it has hungry fledglings in the nest, requiring many hours of hunting to fill their stomachs.

Someone says "shhh," and we hold our breath, listening intently to a confused murmur drifting down from the sky: migrating birds on their way to Canadian forests, perhaps even to the Arctic tundra. Without competition from traffic and other noises of civilization, their voices are much more audible than usual. We can identify the liquid note of the Swainson's thrush, the sharp clink of the white-crowned sparrow, the emphatic notes of flycatchers. But the majority of calls are the feeble chips of warblers, which only an expert could pick out. If we had a telescope focused on the moon, we could see these flocks silhouetted against it as they hurry northward. Again we are awed by the great mystery of migration. How can birds as tiny as warblers, less than six inches from beak to tail, find their way from a swampy forest in Central or South America to the same bit of Canadian wilderness where they first saw light?

The last of the shorebirds have not yet left for the Arctic, but they are restless, gathering into flocks for their northward trek. At intervals we hear a group sweep by on swiftly beating wings. Their calls are easily identified and we can pick out the conversational chatter of semipalmated sandpipers, the pure minor notes of black-bellied plover, and the harsh voices of ruddy turnstones. A late, bewildered lesser yellowlegs whistles loudly.

Suddenly beside us, as though responding to a cue, there is a demoniac squalling, hissing, screaming, and screeching, accompanied by a furious beating of wings. Instinctively we crouch, hair bristling, breath coming in gasps. Gradually, we come to realize that we have only disturbed at their dinner table two of the largest marsh birds: great blue herons.

Instinctively we feel a strong kinship with the Scot who, centuries ago, wrote:

> From ghoulies and ghosties and long-leggety beasties
> And things that go bump in the night,
> Good Lord, deliver us.

June 1 to 30

Multiplication

With few exceptions, all birds, mammals, reptiles, amphibians, and insects are raising families. Behind a sheltering wall of reeds and cattails, Canada geese, ducks, coots, moorhens, rails, and grebes are guarding groups of youngsters. American and least bitterns, whose fledglings develop more slowly, avoid leading enemies to their nests in the tall vegetation. Green-backed herons perch beside their crude nests, built in trees at the water's edge.

Every kind of habitat in and bordering the marshes now holds its quota of nesting songbirds. In the cattails and other heavy marsh growth red-winged blackbirds reign, although a few marsh wrens and, rarely, a swamp sparrow can also be found there.

The border thickets are favored by many birds, including the yellow-billed and black-billed cuckoos, ruby-throated hummingbird, kingbird, willow flycatcher, tree swallow, house and Carolina wrens, catbird, brown thrasher, robin, wood thrush, cedar waxwing, starling, prothonotary and yellow warblers, yellowthroat, red-eyed and warbling vireos, grackle, cardinal, indigo bunting, goldfinch, towhee, and field and song sparrows. Female brown-headed cowbirds patrol the area searching for possible hosts. They lay their eggs in other birds' nests, usually those of a smaller species, and depend on these dupes to hatch and care for their fledglings. Strangely enough, the victims of this stratagem do that very thing.

The few patches of swamp forest shelter a distinctive group: the tufted titmouse, great crested and Acadian flycatchers, wood pewee, northern (Baltimore) and orchard orioles, American redstart, flicker, red-headed and downy woodpeckers, and occasionally red-bellied and hairy woodpeckers. The adjoining fields are homes for eastern and western meadowlarks, the bobolinks, the field, Savannah, and vesper sparrows, an occasional dickcissel or grasshopper sparrow, and the ever-present red-wing blackbird. A few ring-necked pheasants raise their chicks along the grass-grown edges. By the end of the month some species of smaller birds are already busy with second broods.

About 130 varieties nest or have nested in the Lake Erie marsh ecosystem. The major breeding ground of the large waders, however, is on West Sister Island. An estimated 1,200 pairs of black-crowned night herons, 1,000 pairs of great blue herons, and 1,000 pairs of great egrets nest in the large heronry there. Many of these birds make a daily nine-mile round trip to the marshes for food for their nestlings. The few small heronries of black-crowned night herons and great blue herons in the Lake Erie marshes have mostly been abandoned, but large groups of great blue herons have nested for many years in the Sandusky Bay marshes.

Not all the birds that can be seen are breeding. Occasionally a few shorebirds or waterfowl fail to complete their northward journey, usually because they are crippled or diseased. During the month of June, observers who come upon birds of any species that are a great distance from their normal breeding grounds should not jump to the conclusion that they are nesting here.

In early summer, the many miles of dikes become a vast nursery. Burrows are as far apart as possible, mainly for reasons of security, since members of the community prey on each other. One of the numerous dike dwellers is the honest woodchuck, a solid, stolid citizen and an excellent artisan. Woodchucks not only build homes for themselves, but their abandoned dens are used by skunks, raccoons, opossums, and, as a safety cellar, by rabbits. Usually there are two entrances to a woodchuck burrow, which may be seven feet deep and thirty feet long, and the occupant is seldom far from one of them. When alarmed, the woodchuck whistles sharply and dives into its retreat.

The woodchuck is a vegetarian with few bad habits except for undermining banks and dikes with its large burrows. Because of its size, its only major enemies are foxes and human beings, but when the five youngsters emerge after a month in the den, they will encounter many more. And even though parents keep a watchful eye, some will be lost.

At the opposite end of the personality scale is the mink, the perfect predator, so efficient it can run, swim, dive and even climb trees in pursuit of its prey. It eats just about every living thing it can capture and sometimes kills more than it can eat. I once watched a mink dive under water again and again, each time coming up with a crawfish. It would take one or two small bites out of the back of each and discard the remainder. The female mink is a good mother and carefully rears the half-dozen playful pups that leave the den at this season. They appear very innocent, giving no indication of their future terrorist activities.

One dike dweller that is safe from the mink is the skunk, whose chemical weaponry has been perfected through the ages. Perhaps that is why it is not overly ambitious. It is able to dig but prefers a ready-made burrow. As for food, it is not fussy, munching on beetles, black crickets, and the large marsh grasshoppers, to which it adds mice, shrews, frogs, and probably snakes. Its litter of four to ten kittens leaves the den in early summer. They are very playful and attractive, and often accompany their mother on a hunting expedition. Hikers who encounter such a family group quickly retreat to the maximum distance. Probably the skunk's chief enemy is the great horned owl, which apparently has no sense of smell.

For many, the most interesting and appealing resident of the marsh is the raccoon. Its black mask and serio-comic behavior as it traverses the dikes have made it a favorite. Yet like all predators it has blotches on its record—at least from the birder's point of view: raccoons prefer to nest in tree cavities and at one time frequently used wooden wood duck boxes, enlarging the openings. Now that duck boxes are made of sheet metal, and forests beside the marshes are greatly reduced, many raccoons have taken over holes dug in the dikes by woodchucks.

The average raccoon litter is four, and the friendly, playful youngsters begin to leave the den at ten weeks. People sometimes capture them for pets, but this is illegal and dangerous, since they don't remain harmless or friendly for long. Raccoon families find plenty to eat: berries, small mammals, fish, frogs, crawfish, reptile and bird eggs, snails, and even mussels.

Next to human beings, raccoons are the chief enemy of snapping turtles, a relationship that creates an interesting ecological pattern: turtles eat ducklings, and raccoons cut down the turtle hatch by eating the eggs, thereby indirectly helping waterfowl production. But raccoons also devour all the duck eggs they can find, and their climbing skills allow them to raid wood duck nests. To hold both to reasonable limits, turtles are trapped for food in summer and raccoons for fur in winter.

Of all the marsh mammals, the red fox is considered the cleverest at avoiding humans. It finds the dikes to be excellent den sites, though it ranges far more widely than other marsh residents. The conscientious vixen keeps a very watchful eye on the kits playing in front of the entrance to their home. The widespread survival of foxes can be traced to their ability to avoid traps and to their omnivorous habits. Although they prefer mice, they will settle for fruits, frogs, snakes, and insects. They are extremely clever at stalking ducks and pheasants on the nest and at capturing young mammals, particularly stray woodchucks or rabbits. Since ancient times, the fox has been portrayed as a serious enemy of wildlife, even though it seldom weighs even fifteen pounds.

The opossum represents the marsupials, the oldest mammal family now living in the United States. Yet it occupies an across-the-tracks status among marsh mammals and is a Johnny-come-lately on the dikes, having moved northward from southern Ohio around 1920 as the forests were leveled. Its living habits are slovenly, its daily menu abominable. If the size of the brain case is a valid criterion of intelligence, it is far more stupid than its neighbors. Could it be that stupidity has helped it survive?

Opossums really do "play 'possum." Often when an opossum is attacked, it goes limp, opens its mouth in a grotesque grin, and pretends to be dead. Even if moved about or lifted by the tail, it will not respond. Just why this action prevents larger mammals from eating it is hard to say.

One reason for the opossum's persistence through the ages is its power of reproduction: a female may have as many as fifteen offspring. The newborns, as small as honeybees, crawl into their mother's pouch for a five-week growing period. When they emerge, they may cluster on her back for a ride, clinging to her fur or long rat-like tail, a picture few hikers have ever seen.

An opossum is an even less finicky eater than a skunk. Anything will serve, animal or vegetable: fruits, earthworms, frogs, snakes, mice, shrews, or insects. It differs from its neighbors on the dikes in its willingness to eat carrion, even large farm animals in advanced stages of decay.

By far the most numerous marsh mammal is the muskrat. By June, first litters numbering four to nine are out and about, and a second litter no doubt is on the way. Although it prefers to build its home of vegetation in the shallows, it sometimes burrows into the dikes near water level. Passersby can see it swimming, diving, or running on land since it is perfectly at home in both habitats.

The "musquash," as the Indians called it, has long been prized for both food and fur. French pioneers considered it the most important animal in the marsh, and the Catholic clergy even permitted muskrats to be eaten on Friday. The muskrat has many enemies, especially in June when eagles, hawks, owls, minks, foxes, and many other predators are hard-pressed to find food for their own young. The huge Norway rat is a perennial threat. The muskrat holds its own simply by out-breeding its enemies, usually producing three litters a year. It has little difficulty finding food: in addition to the superabundant cattail, it feeds on roots, grasses and similar aquatic plants, as well as clams, frogs, fishes, and other animals. By stabilizing water levels, modern marsh management has helped the muskrat to thrive.

The largest mammal in the marsh habitat is the white-tailed deer. Welcomed back about 1940 after an absence of five decades, it too is busy rearing a family. The drier fields of blue-joint grass or cane are ideal nurseries, where single or twin fawns can find shelter during the day while the doe feeds in the neighboring thickets. A hiker often comes upon a place where the tall grass and cane have been flattened by family groups. These and numerous hoof prints prove the deer are holding their own; over five hundred live in the marshes now.

One morning, as I watched a few shorebirds in the Ottawa Refuge, I suddenly discovered a doe and triplet fawns watching me. When I moved to photograph them, mama doe stamped and snorted in an effort to warn her youngsters away, but they insisted on lingering to observe the strange creature. Finally, they listened to her warning and followed her back toward the woods.

More color has appeared in the marshes. One of our most impressive wildflowers, the white water lily, is in bloom, as are gold spatterdock, blue pickerel weed, white-petaled arrow leaf, and lavender flowering rush. There are still lingering traces of purple iris. In the open ponds and canals are several duckweeds, two pond weeds, crowfoot, bladderwort, and water milfoil. Many rushes and grasses, and over fifty types of sedges, compete with the broad-leaved and narrow-leaved cattails. Wild rice, which was abundant until marsh controls became a necessity, is now gone. In some areas, swamp loosestrife and buttonbush, with its quaint flowering globes, form almost impenetrable tangles. Purple loosestrife, introduced from England, has proven to be a serious nuisance, crowding out vegetation more favorable to waterfowl.

Unmowed dikes are now made virtually impassible by a barrier of vegetation that includes sumac, two species of dogwood, saplings of several trees, and masses of nettles and Canada thistles—all of which make the dikes shelterbelts for many forms of wildlife.

It's June, Let's Visit the Marsh

It's an old town. Situated on the shores of Lake Erie, it was a thriving community thousands of years ago—more ancient than those man-made cities excavated by archaeologists. On a June day, events there are much the same as they have been for centuries, even though a nuclear power plant cooling tower looms in the distance and television towers pierce the sky. The distance separating the very old from the very new is short—only a few hundred feet of cattails or cane rippling in the wind. This is the barrier that has preserved the antiquity of cattail village.

Nowadays, when the only access to the marshes is the network of dikes that contain and protect them, tangled brush makes visits virtually impossible after the end of May. In past years, however, the network of canals built by the early marsh owners furnished access to nearly every part of the marshes, usually by means of a flat, squat, plain, no-nonsense boat called a punt.

Although most marsh inhabitants can be seen from dikes, these structures, essential as they are, nevertheless impose a kind of barrier between the two worlds. In a punt, you are not on the edge of the marsh world but *in* it; moving easily and almost soundlessly along the waterways, you can sample the busy life of an entire marsh in a single day. Since the era of the canals is past, let me take you on a punt trip through the marsh in June, representative of many such trips I made through the Cedar Point Marsh between 1928 and 1976.

Our boat has little freeboard—waves are no problem in the marsh—and both ends are pointed to make traveling through vegetation easier. The punter stands erect at the stern and propels the craft with a twelve-foot pole, widened and thickened at its bottom end so it can find firm footing in the mud and also act as a rudder. Some marsh men preferred a pole ending in a blade like a canoe paddle—a matter of choice. A punt is steered simply by twisting your hips and legs in the desired direction, taking care to keep the pole from sticking in the deep mud; if that happens, you have the unhappy choice of letting go and risking the loss of the pole or hanging on and risking a fall overboard. However, with a little practice you learn to keep the pole free.

In places where old channels have become filled by erosion the silt may be ten feet deep, but with a punt we can traverse inch-deep water or even bare mud. (Some of the old marsh men declared that you could spit on the ground and pole a punt over it.) As we move very quietly along the edge of the channel, the citizens of the wetlands, unaccustomed to visitors, are not particularly alarmed until we get too close. Then a male red-winged blackbird flies above, scolding and displaying his epaulets to their limit. As if by magic, up from their nests pop several of his brown-streaked mates, adding their harsh clamor. As we round a point, a great blue heron leaps into the air with a raucous squawk and flaps away on broad wings over the cattails. Ahead of us a small patch of green rushes moves steadily across the quiet water. For an instant we are puzzled. Then we realize it's a muskrat, moving building materials to the site of the tiny haystack that will serve as its winter home.

On an old log beside the open water, turtles are sunning themselves. The polished painted turtles are first to take alarm and plop into the water. The more blasé Blanding's turtles raise their heads, displaying their bright yellow throats, and appear to ponder the situation before they, too, head for safety.

The channel widens, and at the far edge of the open water we see a cluster of ducks. Closer inspection shows that they are drakes—mallards, wood ducks, a black duck, and a pintail—which have joined the "bachelor club," leaving the task of raising young to their mates. Near them is a pair of blue-winged teal who have apparently not yet decided just where and when to nest. As we move deeper into the maze of cattails, it is immediately obvious that reproduction is indeed the leading activity. Coots shepherd their families of fuzzy, red-headed youngsters toward safety. Common moorhens, with many hen-like clucks, persuade their little black chicks to take shelter in the protecting cattails. Odd calls reveal the presence of king and Virginia rails with their black offspring. The young respond readily to their parents' urgings and disappear almost with the wave of our binoculars.

Silent but striking in its orange, cream, and black plumage is the least bittern, which flies lackadaisically over the cattails. This species has become quite rare. Our watery path takes us beside a small island where, high on a poplar, is a wood duck box. Artificial nesting sites are one of the few innovations cattail village has accepted from civilization. But even these only replace cavities in the dead trees that loggers and high lake levels have eliminated.

As we draw near, a female emerges from the box. Apparently it contains eggs or young. The question immediately arises: How do the ducklings get safely to the water from that height? At one time it was thought that the mother carried them between her feet, and some observers actually claimed to have seen this performance. But now we know that the youngsters simply pop out of the nest and flutter down to safety. In one case on record, nine ducklings hatched sixty feet up an oak stub in a Toledo suburb. When the time was right, they jumped safely to the ground and immediately made their way to a backyard wading pool.

A few land birds live on the island, and as we glide by we can hear the trill of a song sparrow, the staccato calls of a willow flycatcher, and the brilliant songs of a yellow warbler and an indigo bunting. A female mallard, almost surrounded by her large family, moves rapidly ahead of us. One youngster persists in falling behind, and we wonder how long it will survive. Ducklings have many natural enemies—marsh hawks, fox snakes, snapping turtles, raccoons, mink, and even bullfrogs—and so we remain quiet to give them plenty of time to hide. Our caution and slow movement pay dividends when, as we turn a corner, we see a great egret fishing. Slowly it moves through the shallows, yellow eyes fixed intently on the water. Then in a white flash it strikes, capturing a six-inch carp, which it quickly turns around and swallows head first. We can almost see the fish squirming and thrashing as it slides down the long neck.

Suddenly we become aware of frogs. We have been so intent on birds and larger creatures we almost have overlooked them. But they too are dwellers in cattail village. Tiny cricket frogs pipe beside our path, and chorus, leopard, and green frogs leap for safety ahead of us with complaining cries. A pair of bulbous, protruding eyes at the far side of the waterway reveals a bullfrog. At intervals in the distance we can hear the booming, eerie voice of its fellows. Looking for frogs is a large water snake, so old it has lost its reddish brown-and-black checks and now is jet black. It swims calmly away, with only its head above water like the periscope of a submarine.

There is a sudden, violent thumping on the bottom of our punt and several seconds elapse before we realize the cause. It's nothing dangerous, only an oversized carp that we have run over as it was browsing on submerged vegetation. Fishes have been few on our journey, only an occasional showy goldfish or a black cloud of tiny bullheads drifting along in the shallows. But before the marsh was completely diked, conditions were very different: largemouth black bass occasionally leaped entirely over the prow, crappies snapped at insects a few feet away, and savage, slow-moving northern pike and bowfins, always hungry, searched for prey.

High stone dikes now keep out the many Lake Erie fishes that formerly moved into shallow water to spawn. The loss of these fish and the introduction of carp and goldfish in the late nineteenth century constitute the greatest change in the marsh complex and the most serious downgrading of its environment. Carp and gold-fish feed on the vegetation that would otherwise naturally clarify the pond waters, and make matters worse by stirring up bottom sediments that further muddy the water.

Tree swallows sweeping over the open water indicate that gnats are numerous, and soon we begin to notice other insects. Because of their size, shape, and rapid flight, dragonflies are the most conspicuous. These seemingly fragile creatures are among the most ancient insects on this planet. On the plants along the shore we can see the brown cases left when the larvae crawled up from their homes in the water and mud, split open the cases, and emerged as winged creatures capable of flying all over the marsh in a single day. Years ago they were called "devil's darning needles," and every youngster knew they would sew your mouth shut if you talked too much. It is reasonable to suppose that some frantic father, fishing with his son in one of the canals, first invented this tale when questions came too fast.

Now and then we catch sight of an iridescent blue damselfly. Occasionally one of the larger butterflies—red admiral, viceroy, or painted lady—flutters across the marsh. As we enter another open pond, we are startled by several black birds with gray wings that dive-bomb us repeatedly—fortunately hitting only our hats. They are black terns, and their actions indicate that we have encroached upon their private portion of the wetlands. A short search reveals a nest of four eggs laid on a raft of floating vegetation. We move on quickly in order not to disturb them too much, since they are becoming quite rare in the Lake Erie marshes. But we do not overlook the fact that much of the pond is overlaid with the great green platters of American lotus leaves. Unlike water lily leaves, many of these are elevated above the surface of the water. A few buds are visible, a promise of great beauty later in the summer, when these largest of all local wildflowers will bloom.

Now we come upon a mud flat where passing creatures have left their imprint: coots, moorhens, and great blue herons. A set of tracks with a dragging tail means a muskrat has passed, and that great broad trail with deep footprints must have been made by a large snapping turtle. Apparently it was heading for a nesting site on an old dike, where it intended to dig a hole and deposit several dozen eggs. But the hand-like prints of a raccoon heading in the same direction indicate that the eggs probably ended up in its stomach.

A great chocolate-marked, cream-colored fox snake meanders across the channel, its red-brown head raised well above the water. Ignoring us, it moves sinuously toward a colony of red-winged blackbirds. They scream and strike at it, but like fate it glides to the nearest nest, encircles the supporting cattails, and devours the fledglings with appropriate deliberation. We are reminded again that the marsh is beautiful but not gentle, and death is an instrument of survival.

The channel deepens and we surprise a pied-billed grebe with her family, the striped heads of the chicks giving them a distinct, reptilian appearance. Instead of diving, the mother remains in view to attract our attention. But the fledglings, scarcely out of the nest, dive for many yards and swim rapidly away. When one youngster fails to keep up, its mother promptly gives it a lift on her back. Soon they are far ahead of us and safe from fears. A loud "kow-kow-kow" from the father nearby seems to celebrate their successful escape.

Wrens have always been known as good singers and the harsh, gurgling voices of a small colony of marsh wrens comes as some-something of a shock. We recall the ornithologist Dr. Lynds Jones's tongue-in-cheek explanation: "Marsh wrens have harsh voices because they get their feet wet so often." We can see several large globular nests, but we do not bother to investigate them, since the male marsh wren builds several dummy nests and finding the real one is difficult. Whether this is simply overachievement or a sophisticated attempt to lead predators astray is anyone's guess.

We move on quietly and deliberately through cattail village, taking advantage of every bend in the channel and every clump of cattails to shield our travels. Suddenly we hear a weird, wild, outlandish, guttural call unlike any other sound in nature. We approach cautiously and see an American bittern proclaiming everlasting love to his mate. Normally he wears a discreet brown, black, and buff costume, but now he flourishes a white ruff on both sides of his shoulders and neck, and his throat is distended to show a patch of orange skin, as though he had swallowed a rubber ball. He snaps his head back and forth, clicking his beak each time—evidently a bittern version of ". . . and a-one, and a-two, and a-three." He puffs his chest to its limit and launches into his aria with more mugging than an amateur basso rendering "Asleep in the Deep." His head jerks emphatically with each syllable, and his entire body moves with violent, convulsive motions as though he were in great pain.

Then comes his song, an explosive, "gunk-ga-dunk." Two, three, even four times he sounds forth and then collapses, apparently from sheer exhaustion. At no time has our lovesick swain appeared to be enjoying himself. His vocal efforts have been variously described by other listeners as "pump-er-lunk," "oong-kga-chunk," and by New England birders—who are notoriously romantic—as "plum-puddin." The musical quality of the call can perhaps be illustrated by the fact that when I first heard it in my youth I thought someone was trying to start an asthmatic one-cylinder gas engine used to operate a pump. Perhaps the bittern's unusual call has developed because, unlike most of the heron family, he leads a solitary life. He has to broadcast his location to that lonesome female waiting somewhere in the many square miles of cattail and grass, or they might never get together. The bird's built-in amplifying system plays an important part in communicating that he is unattached and available.

Now the performer finally becomes aware of our presence. He collapses his inflated breast, straightens his neck upward to its full length, points his beak toward the sky, and freezes. His stance and coloration resemble the background of marsh vegetation so closely that he almost disappears. But now a soft breeze springs up, swaying the sawgrass and sweet flag on the channel border. As though on cue, the bittern begins to sway too, moving his head and neck sideways in an effort to match the movement of the grasses. But, although he remains sublimely unaware of his mistake, his movements are out of sync with the waving grasses; the poor fellow's intentions are good, but his sense of rhythm is terrible.

Our attention is distracted by a dark shadow sweeping over the marsh. Looking upward, we can see the cause—a magnificent bald eagle with eight-foot wingspread circling above us, no doubt searching for a carp or muskrat. Its white head and tail flash in the fading sunlight. Ten years ago, we might have ended our trip through the marsh on a sad note, ruminating on the imminent disappearance of this magnificent raptor, which was incapable of hatching young from its DDT-contaminated eggs. Today, the picture is different; thanks to better control of chemical wastes and to state fostering programs, the eagles are returning to the marshes.

Evening is coming on, and the mosquitoes are becoming impossible to ignore. As we pole up to the the dock, I wonder, as I have so often, what has been the reaction of the marsh residents to our intrusion. I doubt they were as pleased to see us as we were to see them—but were they perhaps as curious?

July 1 to 31

Hide, Brother, Hide

This is the season when dapper drakes become drab and dismal. All waterfowl look their best in the mating season, but after the hens begin incubating eggs the drakes leave them, gathering together in a sort of stag club. There they gradually undergo a complete molt, even dropping their flight feathers. For nearly a month, usually from mid-July to mid-August, they remain unable to fly and are very vulnerable to enemies. They are feverish, feed but little, and spend most of their time skulking in the cattails, reeds, and grasses. These changes are most noticeable in the surface-feeding species.

To help conceal the drakes through this period, their appearance changes markedly; the drab new garb, known as eclipse or hiding plumage, resembles that of the hens. The male mallard, for instance, loses his bright green head and chestnut-and-gray breast and looks as if he has pulled on a brown coverall. Gaudy pintails, wigeon, shovelers, and blue-winged teal also become brown from head to toe.

In August, when their flight feathers are restored, the drakes begin another molt, which may last up to three months. This time all their feathers are replaced except the new flight pinions, and the birds appear again in their familiar mating plumage. A few species, notably blue-winged teal and shovelers, change so slowly that they won't reach their final stage until next spring.

Black ducks molt, too, but undergo no outstanding color changes, wearing the same sober costume throughout the year. Divers, which are rare breeders in Lake Erie marshes, undergo only a partial eclipse molt, but they do lose all flight feathers. The ruddy duck is an exception. Because he helps raise the "kids," his complete molt is delayed until August and September, and he carries this duller plumage until the start of courtship in spring.

Meanwhile, back at the nursery, the females are too busy to molt. During the winter and spring they go through an inconspicuous change of body and tail feathers and acquire a special coat of long, soft down on their breasts to assist incubation. As soon as the ducklings can fare for themselves, in seven or eight weeks, the hens undergo a complete molt, including flight feathers. They will wear this new outfit through the winter. The fact that all ducks go through a period when they are flightless emphasizes the need for wetlands with heavy cover to protect the birds from their enemies.

Because of the changed appearance of the drakes and the presence of juveniles, ducks are difficult to identify all through the summer and into the fall. This difficulty affects others besides birders: modern duck hunting regulations require a shooter to be able to recognize all species of waterfowl, and many species are still in eclipse plumage during the first part of the hunting season. The best field marks are the wing patterns, supplemented by the shape of a bird and its pattern of flight.

Bird migration in the Lake Erie marsh areas, which ceased when the last of the shorebirds headed north about June 15, resumes during the first week of July when the first shorebirds arrive on their return trip to South America. Thirty-one species are usually seen, each of which arrives and leaves according to its own schedule. The earliest arrivals include upland, least, and pectoral sandpipers, greater and lesser yellowlegs, short-billed dowitchers, and stilt sandpipers. These last-named birds are very rare in spring, since they follow a different route northward. All of these birds are still in breeding plumage, but day by day they change to their winter costumes. By the end of the month, about half of all the species of shorebirds will be on hand.

Starlings, red-winged blackbirds, and grackles now gather together at night in large flocks in the marshes. During the day they forage in the countryside, sometimes traveling many miles. In former years, when they were more numerous, bobolinks joined these flocks, which sometimes reached ten thousand by the end of July.

For most of the marsh dwellers, July is a quiet period when juveniles of all kinds are growing up and beginning to strike out for themselves. Their elders are resting from parental duties as they wait for the beginning of the next breeding cycle.

One of the most fascinating creatures of the summer marsh is the dragonfly, an iridescent beauty with an exceptionally large appetite and the hunting skills to go with it. It can change direction at speeds up to fifty miles an hour, and once its jaws and forelegs clamp onto a mosquito, fly, gnat, or even a small moth or butterfly, that insect doesn't stand a chance. More than once I have lingered to watch this predator, marveling at the skill that makes even a peregrine falcon look clumsy.

At least fifteen species have been found beside Lake Erie. One reason for their exceptional hunting ability may be that they have had so much time to practice: remains have been found in limestone formations three hundred million years old, a long time to undergo natural selection. They were the "birds" of that Paleozoic era, with wingspreads of thirty inches. In addition to speed, the dragonfly's most valuable asset is its eyes: each one contains twenty thousand sight units, or facets, and covers the entire side of the head—equipment a birder might envy.

The family Odonata is divided into the larger, faster, stronger dragonflies and the smaller, more delicate damselflies. Some common dragonflies are the red skimmer, green darner, blue darner, ten-spot, white-tailed skimmer, and the dainty widow skimmer. Among the damselflies, the bluet is by far the most common, but the black and ruby-spotted types also appear in the marshes. The two families have quite different ways of perching. The dragonfly extends its wings out at right angles to its body so that it lies flat and is very visible. The damselfly folds its wings together above its body and extends them backward. Sometimes damsels appear in great numbers. A few years ago on a small area of the Magee bird trail, at least one thousand of them gathered on the ground, for no reason that I could discover. They seemed almost dormant, and I had to walk very slowly through the group to avoid trampling on them.

When mating time arrives, pairs of dragonflies fly about joined together, the male gripping the female by the back of her neck with special clamps at the end of his abdomen. He transfers his sperm capsule to a portion of his body where his mate can reach it and apply its contents to her eggs.

Although their great speed permits them to explore the countryside for food, dragonflies do need water to reproduce. Puddles or ditches that dry up in midsummer are worthless, but the marshes are ideal. Methods of laying eggs vary: some are inserted in mud or decaying vegetation, some into living water plants, and some, encased in gelatin, are dropped into the water. More than once I have watched a female fly back and forth over a pond, dropping down at intervals to deposit an egg.

When they hatch, the nymphs burrow into the mud or crawl about looking like horror-movie monsters in miniature. Like adult dragonflies, they are capable predators with strong jaws, and virtually every smaller creature they encounter, even tiny fishes, be-

A Wetlands Almanac 53

comes food. They themselves are food for the larger fishes, all of the herons, waterfowl, and some of the shorebirds. At the proper time, the survivors drag their ugly bodies up the stem of a water plant, break open their carapaces, and—as happens so often in nature—become beautiful.

One quiet summer day, as I watched a dragonfly perform its daily routine, I found myself wondering how this small creature survived earthly changes for millions of years while giant lizards, great mammals, and Neanderthals all perished. When I thought about it, the reason seemed obvious: the dragonfly feeds on a variety of small insects, which have always been available, and its rapid, strong flight has enabled it to find distant food and water as well as to escape glacial advances or droughts. In a word, it is *adaptable*. In our day, its greatest enemies in the adult stage are the kestrel, kingbird, crested flycatcher, phoebe, black tern, all of the herons, various waterfowl, and even bullfrogs.

Considering its huge appetite for insects, many of which are either a nuisance to humans or carriers of disease, it seems to me that we don't appreciate the dragonfly as much as we should. Butterflies get far more attention.

Incidentally, the only insect older than the dragonfly is the cockroach.

August 1 to 31

From Arctic to Argentine

In August the marshes reach their peak of color and inaccessibility. Cattails, reeds and grasses are at their highest, and the ponds and shallower canals are choked with aquatic plants. The borders are thick with shrubs, tall nettles, swamp thistles, and vines: burcucumber, moonseed, the two bindweeds, virgin's bower, poison ivy, Virginia creeper, dodder, and wild grape. Unmowed dikes are, if possible, even more overgrown than in June. Only the swamp forests, in which the ground cover is minimal, are still passable.

The two largest flowers in northwestern Ohio are now in bloom. The pink swamp rose-mallow, with blossoms about five inches in diameter, is both showy and abundant. The lotus, with leaves up to two feet across and eight-inch yellow flowers, is less common. These two are in marked contrast to the smallest plants, the duck weeds, which measure less than one inch in length when full grown, yet are so abundant that they cover acres of water.

The purple hedge nettle is a very prominent mint. Other blue and purple flowers include nightshade, purple loosestrife, common and swamp milkweeds, bellflower, large lobelia, ironweed, vervain, two varieties of skullcap, monkeyflower, wood-sage, and swamp thistle. Yellows are furnished by goose tansy (silverweed), evening primrose, coneflower, sneezeweed, bur marigold, and several kinds of wild sunflowers. White accents are added by yarrow, boneset, Queen Anne's lace, water lily, arrow leaf, and the various forms of water hemlock.

Two of the most beautiful and unusual blooms are the pink-petaled boltonia and the false dragonhead. About fifteen forms of polygonum and both yellow and orange touch-me-nots can be found. Adding contrast to the picture are the seed clusters of the many grasses, rushes, and sedges and the newly-formed cigar-shaped brown spikes of cattails.

A constant stream of shorebirds passes through the marshes in August, assuming winter plumage. Some of the more unusual types are the avocet, the willet, the two godwits, the buff-breasted and western sandpipers, and the red-necked and Wilson's phalaropes. The rare red phalarope arrives late in the season. Birdwatchers journey many miles to observe these concentrations of shorebirds. There are several reasons for this popularity: shorebirds are among the world's greatest travelers; more than thirty-five different varieties, including some very rare ones, have been noted here; their habitat is limited to mudflats and sand beaches; their identification poses a challenge, even to the expert; and one never knows exactly where to find them or whether a rare species will be discovered.

As the shorebirds move over their feeding grounds, they gradually segregate according to species—not from any sense of kinship, but because of habitat. Some prefer the shelter of the cattail border, some the mud dotted with small clumps of vegetation, some the water's edge, and some the shallows.

The killdeer is one of the birder's greatest handicaps. This self-appointed watchman is always alert to the presence of humans and sees potential danger in everyone. Worse, its loud warnings are heeded by all other shorebirds. Most species will permit a birder to approach fairly closely, at least on occasion, but not the killdeer. You may creep toward a flock to verify an identification or to snap a picture, using every bit of scanty cover available. Then, just as you reach a good vantage point, the killdeer sounds the alarm and away goes the entire group like a single bird.

Shorebirds have a more pronounced flocking instinct than any other bird family. They fly very rapidly in a compact unit that darts and swerves over the marshes, changing course with no hesitation and at no apparent signal. As they perform their intricate aerial maneuvers, their bright white underparts alternate with the brown of their backs like the intermittent flashes of a signal light. So perfect are their movements that they give the impression of a large organism governed by a single brain. Sometimes the birds continue their flight until they are no longer visible, but normally after a few sweeps they drop suddenly to earth and nonchalantly begin to feed. This alarm flight may occur several times within an hour, often for no apparent reason.

On one occasion, a bald eagle was wheeling over the marsh, doing me a service as it did so. For wherever it went, the birds of the marsh—ducks, herons, and moorhens as well as shorebirds—would rise up, not so much in terror as from force of habit, giving me a good look at them. I was amused to note that when the eagle finally alighted on a muskrat house nearby, the ducks—even the pint-sized blue-winged teal—fed unconcerned within a few dozen feet. Even when the eagle departed, scarcely a ripple of excitement passed over the feeding birds.

But suddenly, they scattered like leaves in a November gale. Black ducks, mallards, blue-winged teal, wigeon, and pintails exploded from the shallows. Moorhens and coots pattered to the security of the cattails. Yellowlegs screamed and fled. Even that awkward giant, the great blue heron, left hurriedly. Then from the blue sky dove the cause of the alarm, the streamlined peregrine falcon, warrior of the skies. Just a fraction of the eagle's bulk, it nevertheless carries death in its huge talons. And the birds of the marsh knew it.

Although the peregrine once was called a duck hawk, it seems to prefer shorebirds when they are available. The chase is usually not very long. If the surprise attack is not successful, the hawk gives up and returns to its perch. At times it appears to be driving the smaller birds as though it were playing a game. But shorebirds must have short memories, because they soon resume feeding.

Near Cedar Point Marsh I once witnessed a prolonged skirmish between a peregrine falcon and a single sanderling. This particular shorebird had been cut off from its companions on the beach and driven out over open water. Time and again the hawk tried to get under its prey, to keep it off the water, but time and again the sanderling darted downward to the surface of the lake. Occasionally it appeared to dive under water. More than once it escaped by a hair (or should I say a feather?), as speed and agility overcame the predator's superior power. How long the pursuit went on I can't say; I was too engrossed in the contest to think in terms of minutes. But finally the falcon must have decided that such a small bird wasn't worth all the effort. It gave up the attack and the sanderling rejoined its fellows.

Herons hatched on West Sister Island or in southern swamps often visit the Lake Erie marshes in late summer. These are the snowy egret, the little blue heron in immature white plumage, the cattle egret, the glossy ibis, the rare tri-colored heron and, in past years, the greater flamingo. Even the largest birds in eastern North America occasionally visit our wetlands: white pelicans, with their enormous bills and eight- or nine-foot wingspread, and, rarely, the smaller brown pelican.

Puddle ducks from the north begin their southward migration in mid-August, and by the final week of the month as many as twenty thousand have been found in the Ottawa National Wildlife Refuge. Many of these migrants, as well as breeding blue-winged teal, appear to depart for the south long before cold weather arrives. Bobolinks and orchard orioles also leave before September 1, and the first songbird migrants arrive: olive-sided and yellow-bellied flycatchers and several of the warblers, heralding the great flood of travelers in September.

At times great numbers of nighthawks circle irregularly over the landscape, darting and diving as they feed on insects. Sometimes these loose flocks are spread over several miles. Also attracted by the swarms of gnats, hundreds of swallows feed over the marshes and roost in long lines on the utility wires at its borders. First to flock are the bank and rough-winged swallows, then the barn swallows and purple martins, and finally the tree swallows. Bank and tree swallows are the most abundant, but all varieties appear to be fewer in number than they were several decades ago. During the drought years of the early 1930s, these flocks numbered in the hundreds of thousands.

Most birds are silent at this time of year; if it weren't for the various waterbirds, the marshes would be almost quiet in daylight hours. Among the perching birds, only the coarse voices of blackbirds are noticeable as flocks grow from day to day. But shorebirds, ducks, gulls, terns, and herons never seem to be silent.

August is the month when marsh hikers come to appreciate butterflies. Although waterbirds are plentiful in the marsh, they take care to stay out of sight, and most are in their dull eclipse plumage. Shorebirds are confined almost entirely to sand beaches and mudflats. The fall songbird migration is yet to begin, and virtually all the birds that can be found are in molting or dull plumage. So it is only natural that our attention should turn to the twelve species of large butterflies that brighten the marshes.

Everyone has a favorite butterfly. My own "most beautiful" is the red-spotted purple, especially when its iridescent colors gleam in the sunlight. I also like the less common tiger swallowtail. Except during monarch migrations, the most numerous butterfly in the marshes is the black swallowtail. The oddest name belongs to the painted lady; the oddest shape, to the angle-wing. Two species come early and stay late: the mourning cloak emerges from hibernation in March, and the brighter red admiral follows soon after. The viceroy is chiefly notable for its close resemblance to the larger monarch. The flashiest name—and, for some reason, the hardest for me to remember—is "great spangled fritillary."

Of all the butterflies, the most famous is the beautiful and valiant monarch. While fishing from a boat far out on Lake Erie, I have watched monarchs flutter by, faltering and erratic, looking as if they would drop into the water any second. Compared to birds and even dragonflies, they appear to be poorly equipped for a long journey. In fact, however, monarchs are great travelers: in recent years, thousands have been tagged in northwestern Ohio in an effort to trace their 2,000-mile migration route from Ontario to winter homes in the mountains of central Mexico. At this time of year, their southern migration brings them across Lake Erie and into northwest Ohio. Astonishingly, none of the migrating butterflies has ever made the trip before.

At dusk, monarchs rest from their long trek, gathering by the hundreds on a tree or shrub in a vibrating blanket of orange and brown and black. Some of the wings that pulse in the dying light are marked with paper tags. In the Toledo area, Doris Stifel is the acknowledged expert on tagging monarchs. When she is not checking out bald eagles' nests for the Ohio Division of Wildlife, Doris raises monarchs in her home for release in the wild. She is constantly on the alert for areas where she can gather common milkweed to feed her caterpillars.

On a recent August morning, Doris and I headed for the usual monarch haunts, clumps of trees near the wetlands and fields of red clover or alfalfa. On the way, she was stopped by an engineer at Maumee Bay State Park, who reported a large colony of the butterflies in a small group of trees surrounding a house on Cedar Point Road. When we arrived, the lower branches of the trees were covered with monarchs. Doris picked up her long-handled, lightweight landing net and soon was hard at work.

At the top of one wing near the body, she rubbed a small area clean of scales on both sides. Then she folded a piece of adhesive-coated paper over the bare spot, where its light color was clearly visible against the brown of the wings. The tag sticks firmly without impeding the insect's flight, and bears a printed number and the message, "Report number to Zoology, University of Toronto, Canada."

In a comparatively short time, Doris tagged seventy butterflies. She often scooped up three or four with one sweep of the net, a method of capture that does not harm the insects in any way. The most she ever tagged in one day was 250, in 1991. Her total count up to that time was 30,000, with twenty-five returned tags. This rate seems discouragingly small until it is compared to the average return of one in 10,000. Of the twenty-five returns, four were sent from Mexico. Others revealed a southward flight through Kentucky, Tennessee, Pennsylvania, and Louisiana. The best reward for her many hours of work has been three returns of insects that were all tagged on the same day: one was sent from Kentucky, another from Tennessee, and the third from Indiana. This reveals a very broad migration path, perhaps because the butterflies are easily blown about by strong winds.

As the noon hour approached, the monarchs began to leave, probably in search of food. We decided that, as long as we were near the wetlands, we would take a "bug hike" to see what other insects we could find. Neither of us could be considered an entomologist, but we could surely find some of the larger insects.

On a trail beside a waterway in the Ottawa Wildlife Refuge, we encountered the usual noisy crowd of grasshoppers and crickets, along with many small beetles in a variety of colors and designs. They were mostly unimpressive, except for the stag beetle waving its giant pincers. On one spot, we found a carrion beetle feeding on a dead shrew.

A preying mantis attracted our attention, with its acute-angle legs and heart-shaped head, looking a good deal less savage than it is. Doris recalled that she once spied a monarch perched on a shrub and was about to net it when it vanished. A close-up examination revealed that a mantis had chosen that instant to snatch the butterfly. It had already taken its first bite.

At one point we came upon a circle of about a hundred whirligig beetles, swimming round and round in their three-foot circle. Some were a bit erratic, but most of them held their course with remarkable precision. The sight reminded me of a long-ago bit of philosophy from my brother Bernard, who was an English professor at John Carroll University in Cleveland. We were gazing at a similar group of whirligigs, and I remarked that they must have the most monotonous life of any creature on earth.

Bernard looked at me with a grin and said, "Did it ever occur to you that their routine is ours in miniature? Suppose you selected any city neighborhood and kept track of the comings and goings for a year. What do you think it would look like? Daily trips to work and back, school and back, grocery store and back, spinning around month after month in a giant circle . . ." "Okay, okay," I said, "my apologies to the beetles!"

We were pleased to find a click beetle, nearly two inches long, with its black back and oversized head speckled with white dots. Its outstanding feature is the big, showy, imitation eyes—solid black oblongs emphasized by pronounced white borders. Its real eyes, near its mouth, are almost invisible. Presumably this design evolved as a way to frighten the beetle's enemies. Judging by my own reaction the first time I saw one as a child, this strategy was a great success.

I'm sure that a more careful search, with field guide in hand, would yield dozens more insects, each of them fascinating in its own way. But I have to confess that I get the most pleasure from the most beautiful of the insects, the dragonflies and butterflies.

After Doris left, I headed for the Crane Creek beach to check out the ridges of white shells—reminders that the zebra mussel has arrived in Lake Erie. This Asiatic clam, which looks like our native clams but is only half an inch long, can glue itself in seemingly infinite numbers to any solid surface—a rock, for example, or a metal intake pipe, or a wooden boat. First reported in this area in 1986, by 1988 its population was already enormous, with colonies of up to thirty thousand individual mussels per square meter.

The zebra mussel was introduced to the Americas when large freighters in Asian ports took on water ballast containing millions of tiny mussels in the pre-shell, "veliger" stage. They were dumped into Lake Erie when the ballast was replaced, and they are here to stay. In late spring and summer an adult female can produce up to forty thousand eggs, which hatch within ten days into barely visible veligers. In this form they float through the protective screens on water intake pipes, then cling to the inside surface and grow their shells. Their colonies have clogged pipes in water treatment and power plants all around Lake Erie.

An adult zebra mussel filters a liter of water a day in its quest for phytoplankton, the microscopic plant life it feeds on. This filtration clarifies the lake water noticeably, letting in sunlight and encouraging the growth of aquatic plants—good habitat for walleyes, black bass, northern pike, and even muskellunge. In addition, discarded mussel shells covering muddy areas of the lake bottom could make good nests for spawning walleyes. On the other hand, the zebra mussel's voracious appetite may create a shortage of phytoplankton, which are presently an important element in Lake Erie's food chain. Only time will tell how the zebras will ultimately affect Lake Erie fishes.

In the meantime, the mussels themselves are food for many fishes, including drum, yellow perch, and probably walleyes, white perch, catfish, and bullheads. Diving ducks love them, a passion that probably accounts for the large numbers reported in 1990: 10,000 canvasbacks, 2,500 redheads, and 50,000 common mergansers seen in January and February; 22,600 greater scaup and 46,122 lesser scaup in April; and 234 scoters shot during hunting season in Magee Marsh. The number of greater scaup was most impressive, since previous high counts were in the low hundreds.

The invasion of the zebra mussels is the most recent in a series of environmental events that have dramatically affected the western end of Lake Erie.

- The first such event was the draining of the Black Swamp in the late 1800s. When the huge swamp forest was cut down and the bare soil planted for crops, an estimated two to six million tons of earth washed into Maumee Bay and Lake Erie. The muddy water smothered aquatic vegetation and choked spawning areas with silt, driving away cisco, whitefish, black bass, sunfish, bluegill, rock bass, and northern pike.
- The second event was the introduction of carp in 1881. By 1893, they were abundant in the marshes, muddying the waters and destroying bottom-rooted vegetation.
- Shortly thereafter, sturgeon were eliminated from the lake by

overnetting. In 1881, 531,250 pounds of sturgeon were taken from the Ohio portion of Lake Erie alone. By 1916, they were mostly gone.

• Uncontrolled commercial fishing also destroyed cisco and whitefish. After peaking at over ten million pounds in 1918, the population collapsed.

• In 1921, the parasitic sea lamprey invaded Lake Erie, but for once muddy water was a plus, pushing them northward into the upper Great Lakes. There they became a serious menace until a chemical was developed that destroyed their eggs without harming other fishes. A few lamprey nests were found in Swan Creek in Highland Park, Toledo, but they never were a threat to Lake Erie fish.

• The sixth event, and the first to be welcomed, was the introduction of the smelt. This tasty little fish was introduced into Lake Michigan, then moved into Lake Erie in 1936.

• In the 1950s, the muddy bottom of Lake Erie was poisoned by factory wastes, eliminating the mayfly nymphs. These nymphs, called wigglers, were a favorite bait of anglers. Mayflies once numbered in the millions, furnishing food for all Lake Erie fishes and many birds.

• By the 1960s, commercial fishing had eliminated saugers and blue pike. In 1916, over six million pounds of saugers were netted; in 1966, twenty pounds. Netted blue pike peaked at just under twenty million pounds in 1955, then dropped to fifty-two pounds in 1966 and are now considered extinct in this area.

• Another welcome introduction was that of white perch. These fish migrated to Lake Erie from the Atlantic coast via the Welland Canal. The first two white perch were netted in 1953 in the northeast corner of the lake, and by 1988 they had become common.

• The tenth major environmental event was the first appearance of zebra mussels in 1986. What will be number eleven? Once I would have predicted the elimination of Lake Erie walleyes, but now that they are protected from commercial netting, I can foresee nothing obvious, good or bad. Walleyes, by the way, only managed to survive through a narrow escape: after a 1956 peak of over fifteen million pounds, their collapse in succeeding years would have been disastrous but for the constant restocking of walleyes from Lake St. Clair, by way of the Detroit River.

September 1 to 30

Which Way Is South?

Even if the cicada had not been heard in late July, droning its warning that cold weather is on its way, that message would be carried by the great flood of migrating birds that arrives as the first cold front sweeps the north. This is problem time for observers, the month when birds are the most difficult to identify. Even the most experienced birder is sometimes puzzled, and some even stop their field work until late autumn. Waterfowl are in mixed, obscure plumage, shorebirds have partly changed to winter plumage, and some species that are bright-patterned in spring are now brownish-gray: black-bellied and golden plover, red knot, and dowitcher. Fortunately, about eight species show little change.

Probably no group of birds is as difficult to discover or as puzzling to identify as the warblers. None is singing, some adult males have undergone a complete change in appearance, and females are extra-plain and very shy. Male and female juveniles show great variety in their patterns: sometimes both sexes look like the mother, sometimes young males look like the father, and a few do not resemble either parent. Adding to the confusion are the flycatchers, which are not easy to distinguish even in spring.

About the middle of the month, other varieties of birds begin to appear, notably the two kinglets, the gnatcatcher, winter wren, sapsucker, brown creeper, the three thrushes—Swainson's, gray-cheeked, and hermit, in that order—the rose-breasted grosbeak, junco, and white-throated sparrow. The male indigo bunting is a startling vision in blue and brown as he changes to his new traveling clothes. The male scarlet tanager seems hesitant to display his piebald costume of red, yellow, and black.

This is the time of year when the serious birder should make friends with a bird bander. Relative newcomers to the ornithological scene, banders have contributed enormously to our knowledge of birds. They begin by stringing nylon nets, about forty-two feet long and nine feet high, between slender poles planted in areas frequented by birds. The nets appear delicate, but their thin strands, all but invisible to the birds, are tough enough to entangle any bird that flies into them, without inflicting any harm.

Mark Shieldcastle, a biologist with the Crane Creek Wildlife Experimental Station in Magee Marsh, has been banding for fourteen years. In recent seasons Mark, with his wife and colleague Julie Shieldcastle, has set up nets in the wooded border of Navarre Marsh, which is owned by the Toledo Edison Company and managed by the Ottawa National Wildlife Refuge. The Davis-Besse nuclear power plant and cooling tower occupy the higher ground a quarter of a mile away. Since the loss of the trees at Little Cedar Point, this area has become the major concentration point for migrating birds in the Lake Erie marshes.

Once his nets are in place, Mark inspects each one frequently. He removes each bird with great care, weighs it, measures a wing, records its sex, age, and general health, and fastens a numbered band on one leg before releasing the bird. The entire process takes about two minutes. The data are sent to the U.S. Bird Banding Laboratory in Laurel, Maryland, where they are kept on record. When a bander works in the same area for several years, some birds, particularly breeders, may be recaptured several times.

Often the nets snare too many birds to "process" immediately. For these occasions, banders keep a set of zippered mesh bags, sorting the birds into them by size and disposition to avoid injury—grosbeaks, for instance, are always belligerent and must be kept away from other birds.

Despite the handling, the birds do not appear to panic, although grosbeaks and certain flycatchers will peck savagely at the hand that holds them. During one of my mid-September visits, Mark removed an adult gray-cheeked thrush from a net and handed it to me. I carried it over to a dogwood shrub, loaded with clumps of white berries, to take its picture. While I was focusing my camera, the bird tossed its head back over its shoulder and calmly plucked two or three of the berries, swallowed them, then stretched out its neck for more. After I released it, the thrush stayed on to finish its meal.

Banding provides an excellent picture of bird migration. A birder's information on numbers and species of birds is limited to what can be seen and heard. Vision is impeded by trees, bushes, and long grasses, and by the exceptional ability of some species to remain cleverly hidden in the undergrowth. And in the autumn, birds do not sing. Moreover, a birder may unknowingly count the same bird over and over. Finally, much as we hate to admit it, a quick glance at a bird—especially in autumn—cannot always identify it accurately. None of these problems afflicts the bander, who can identify and count every bird captured. Another advantage, this time from the bird's point of view, is that a rare species needn't become a bird skin in a museum in order to become part of the official record.

Banding makes its greatest contribution to bird identification in the fall. Immature birds in special plumages or adults that bear little resemblance to their appearance in May can be photographed at arm's length, and even the least expensive camera can provide close-ups that were impossible in past years except with injured birds. These photographs and close observations have already resulted in field guides with greatly improved illustrations of obscure plumages.

I asked Mark what was the most unusual bird he'd ever captured, and his reply was startling: "It wasn't a bird, it was a bullfrog!" A yellow warbler had become entangled in a net strung across a shallow bit of marshland, and was hanging head down, just above the water. Its struggles attracted the attention of a large bullfrog, which leaped up and grabbed the helpless bird in its jaws. The bird wouldn't come loose, the frog wouldn't let go, and that's where matters stood when Mark arrived on his regular inspection. Taking in the situation at a glance, he shook the frog loose, then untangled the shaken but unhurt warbler, wiped it off carefully, banded it, and released it. What a story that bird had to tell its progeny!

When I asked if the netted birds were in much danger from predators, he said surprisingly few give any trouble, including the major mammalian predators, red and gray foxes, skunks, opossums, and raccoons. Garter snakes, especially the melanistic (black) ones, sometimes try to reach birds caught near the ground, but without success. Sharp-shinned hawks that dive for trapped birds become entangled themselves and are banded for their trouble.

To my amazement, the animal that does the most damage is the white-tailed deer. Both bucks and does will attempt to chew the captive birds, and can destroy nets by running into them, a mishap I witnessed myself on one occasion. Since repairs are more trouble than they're worth, replacing nets is one of the necessary expenses of banding birds.

As a lifelong birder, I get a real thrill out of holding in my hand a songbird I have seen before only through binoculars. Holding it gently by the legs, I can closely examine areas of the plumage that have always puzzled me. I can look into the large, round eyes and smooth the ruffled feathers. The educational aspect of this scrutiny takes second place to the fact that I am becoming personally acquainted with an individual bird that until now was merely a member of an interesting species.

Fall is a time of rapid changes in bird populations; as some varieties move in, others move on. Numbers of all shorebirds except kill-deers, dunlins, and sanderlings shrink during the month. The long-winged fliers move in, accompanied by the more unusual great black-backed gull, laughing gull, and Franklin's gull. Terns thin out rapidly, but herring, ring-billed, and Bonaparte's gulls stay on until Lake Erie freezes. Herons and great egrets are numerous, the juveniles distinguished more by their lack of fear of people than by their appearance. Soras and moorhens become temporarily common, then depart quietly. Coots arrive and gather together in sociable flocks for a long stay.

Broad-winged hawks drift southward in swirling flocks—sometimes so high they are almost invisible—but their numbers are smaller than in spring. Now the birder watches for the more unusual predators: osprey, merlin, and peregrine falcon. Although the fall migration of all birds is more leisurely than in spring and has few peaks, an observer can list over eighty species a day up to the end of the month.

Adult and juvenile mammals are busy. Muskrats are putting the last touches to winter homes, fox and red squirrels and chipmunks are storing food, rabbits are raising their third litters, and woodchucks are gorging themselves for their long winter sleep. One of the oddities of the marshlands is a color phase of woodchucks in platinum blond fur. Reptiles, frogs, and insects disappear so gradually that their declining numbers are not noticeable until suddenly they are rare.

Further signs of autumn are registered in a changing landscape. A touch of red appears in the leaves of poison ivy, Virginia creeper, sumac, and the dogwoods. Box elder and cottonwood leaves begin to yellow and drop, and here and there splashes of brown contrast with the fading greens of the grass and cattails. Flowers have dwindled to the purple and white masses of wild asters, punctuated by a few lingering goldenrods and sunflowers. Although not as spectacular as spring, September is a period of great activity, as residents prepare to dig in for a long winter and visitors prepare to depart for more southerly winter homes.

October 1 to 31

The Hunting Season

During this month the borders of the marsh gradually change to gold, red, and purple, reaching a peak about the third week. Most colorful are the scarlet of sumacs, dogwood, poison ivy and Virginia creeper, the vermilion of red maple, and the gold of hickories and cottonwood. Last to change are the various oaks, their leaves ranging from bronze to russet. The marshes become more brown each day, in varying shades that mark cattail, cane, grasses, and plants of the open water. Only a few asters, wild sunflowers, and evening primrose are now in bloom.

In October the largest numbers of ducks and geese gather in the marshes on their southward migration, and hunters finally come to the end of their long wait. They have maintained the correct water levels, laboriously planted the various kinds of vegetation that waterfowl prefer, and now they hope to be rewarded.

Some well-meaning people question the advisability of permitting waterfowl to be shot. I can only repeat that no marshes exist between Toledo and Port Clinton which have not been preserved and maintained by hunters or with funds provided by hunters. All of the other marshes have either been drained or filled with refuse. Federal game laws are based on the number of birds available, and declining species are protected. Nor are excessive numbers of waterfowl shot: except in a very few marshes where shooting has been open to the public for a fee (a practice now halted), the total harvest has averaged about one duck per acre per season. If the marshes had been destroyed, the ultimate toll of waterfowl would have been unimaginably greater.

Until 1913, the shooting of waterfowl was permitted in spring. This was not as destructive as it may seem, since access to the marshes was severely limited by the absence of navigable channels. Until the drought years of the 1930s, the season extended from September 16 to December 31, but it was then cut back because of a decline in breeding ducks. In 1941, the season was set at October 1 through November 29, and has varied between those limits ever since. In the last several years, a short season on teal has been permitted. The time of day when shooting began and ended has varied through the years, within a spread from half an hour before sunrise to half an hour after sunset.

The commonest and most popular method of duck hunting is from a blind, because it can be enjoyed by shooters of any age and either sex. Each year in the Magee Marsh and the Ottawa National Wildlife Refuge, a limited number of hunters are chosen by lottery and assigned in pairs to permanent blinds. These are usually built of wood painted dark olive, enclosed on three sides and partly roofed, with a floor raised above the high water mark. The upper half of the front portion is open, and a board across the back provides a seat for what are sometimes very long waits. The entire structure, especially the front, is camouflaged with bundles of cattails or marsh grass. One side is fitted with hinges or straps to provide an easy entrance. In earlier years, some marsh owners substituted metal waterproof tanks sunk into the water and mud, making the hunter much less conspicuous.

Noting the way blinds are scattered through the marsh, casual observers might think they were placed for convenient access. But experienced marsh hunters know that when gale winds drive ducks from Lake Erie, certain ponds are more attractive to the birds than others, and that is what determines where they set their blinds.

As for the number of blinds, the more the better, since there are so many variable conditions to prepare for: natural food is more plentiful in some places, ducks usually come in for a landing against the wind, and some waters freeze more quickly than others. It is the last open place left in a frozen marsh that becomes the perfect location. These observations apply primarily to privately owned marshes and to areas such as Magee Marsh and Ottawa, where blinds are permanent. However, wetlands like Metzger Marsh present an entirely different picture. These are open to all shooters on a first-come, first-hunt basis. Competition is terrific and each hunter must provide his own concealment.

Usually these hunters use a framework on their duck boats, attaching a row of cattails or reeds or an enveloping screen of green-brown netting. Some hunters build a blind in a favorable place and take a chance on getting to it ahead of the competition, sometimes as early as three A.M. Most active hunters are true sportsmen: they remain very quiet and motionless, do not aim over a competitor's territory, and avoid sky-shooting.

Fields where grain has been harvested adjacent to marshes attract large flocks of ducks and geese and, naturally, hunters. Permission to shoot is purchased from the landowner. Some shooters construct a blind from cornstalks or reeds, or they may crouch in a ditch. But most bring in a portable, lightweight blind about three feet by seven, built on a framework of thin-walled aluminum conduit bolted together. Usually it has overhead protection—thin sheets of plywood are most effective. The more experienced shooters, anticipating long waits, build in some sort of seat. Most beginning waterfowl hunters do their first shooting in a field.

One type of duck hunting in a class by itself is open water or layout shooting. This approach, chosen by only the most vigorous and zealous hunters, requires setting up a large fleet of decoys on Lake Erie or one of its bays, crouching down in a small duck boat, and waiting. Results are usually fairly good, but because the boat is in open water, the hunter's life can depend on paying close attention to weather conditions.

I must admit, I have tried layout shooting only once. It was a cold, cloudy day with a stiff north wind, and I lay flat on my back in a small, rocking boat, trying to find the best position for my heavy gun. I missed a few shots. A high wave poured several inches of icy water into the boat. I thought about the relative comfort of a permanent blind, a camouflaged duck boat—even a ditch. After the second wave came, I signaled to the control craft that I'd had enough. In recent years, hunting ducks on Lake Erie has been made much easier by the great stone dikes at Cedar Point Marsh. The hunter reaches a desirable place along the shore by boat, sets out scores of decoys, and hides among the rocks.

The number of species and the total population of ducks vary greatly as the season progresses. All river and pond species are numerous during the last part of September and most of October. Blue-winged teal, shovelers, and wood ducks move out first and more black ducks arrive. The numbers are affected by the fact that mallards, black ducks, and a few pintails and wigeon feed in adjoining grain fields at this time, while most gadwalls, shovelers, wood ducks, and both species of teal remain in the marsh. The latter group provides shooting even on sunny, quiet days when many ducks are not moving. As these more sedentary types leave for the south during October, the hunter has less chance of bagging his limit. Diving ducks seldom visit the marshes in autumn, but may be taken by those who hunt on the lakeshore.

In addition to a suitably constructed and located blind, requirements for a successful shoot include a set of decoys, an artificial caller or the ability to imitate a mallard and a Canada goose, a good retriever, the ability to recognize the various species, and a companion who shoots only at birds within range.

Decoys should present no problem. Any one of the oversized mallard and black duck blocks is excellent. All marsh species are attracted to these, and for some reason the large models have more pulling power. Individual hunters vary in their preference for cast or carved decoys. In areas where pintails are numerous, a few exact replicas of drakes will help entice these wary ducks. In Pintail Marsh I once saw a row of decoys that were real blacks and mallards, mounted by a taxidermist and fastened to a plank. Actually, these "true to life" types did not seem any more successful than carved or cast decoys. As the season progresses, ducks become fewer and more blocks are needed to attract them. Many shooters make a practice of placing the farthest decoy at the limit of effective shooting range.

A caller who can imitate a mallard often means the difference between a successful hunt and failure. Many of the duck's vocalizations can be reproduced, but four are most important: the welcoming "come on in," the coaxer, the feed call, and the unattached-hen call. These can be learned in the wild, but it is simpler to get them from a recording. The calls must be done accurately or waterfowl may become frightened. I once hunted with a partner who must have sounded the alarm instead of the "come and get it," because, as he exclaimed, "I called them ducks and they all flew out to Lake Erie!"

Because all shooting in a marsh is done near dense cover such as cattails, marsh grass, or reeds, a good retriever is needed even when gunners are accurate. Otherwise, depending on the size of the pond hole, winds, and other circumstances, losses may be as high as fifty percent for beginners and twenty-five percent for skilled hunters. The dog most often used is the Labrador retriever. It is large, powerful, easily trained, eager to hunt, and gentle. It can also be taught to hunt ring-necked pheasants, especially in tall vegetation, and even rabbits.

The Chesapeake Bay retriever is probably the best choice for finding ducks in cattails and other very heavy growth. It has an excellent nose, loves the water, and never gives up. For owners who want a combination household pet and hunting dog, the golden retriever is hard to beat, especially if there are children in the home. Several other breeds are acceptable, especially the smaller varieties. Conditions in the marsh where the dogs are to hunt will affect the choice. All dogs should be trained to respond to calls, a whistle, and hand signals to direct them toward a downed bird. One of the biggest challenges is to teach them to remain perfectly quiet until needed.

Of course dogs, like people, have their idiosyncracies. John Ketterer's history of the Toussaint Shooting Club (see p. 150) includes an account of Willie, a retriever with "an irresistible affinity for wooden decoys." This dog would wait patiently until he was ordered to retrieve a duck, then leap into the water and proudly drag in the decoys.

During the last few decades, regulations have required that the hunter be able to recognize the various species and to distinguish males from females. Obviously this is impossible for a hunter who tries to shoot at distant birds. Each season certain species are protected and some are limited to a take of one or two birds. The point system of computing a day's limit calls for absolute identification of every bird shot.

On nearly all hunting days, the first and last hours are most productive, but most hunters prefer to get out early. Somehow, plodding along a trail through the cattails or punting quietly through a channel to a blind at the first sign of dawn has more than a touch of adventure. Ducks can be heard calling in other parts of the marsh and a few are usually flushed when the pond hole is reached. Decoys are placed with whispered instructions and a minimum of splashing. Then comes a period of time-checking until the magic minute arrives. Guns are loaded and there is a tension and excitement not matched at any other time of day.

This feeling does not always last. Hunters have often crouched in a blind and watched several thousand ducks pass high overhead with nary a one coming within gunshot range. This has a tendency to depress the spirit. Once in a while, a small band of Canada geese will cross the marsh at a low level. Then all of the hunter's skill goes into coaxing them to approach the blind. No woodwind player in a symphony orchestra was ever more earnest. Sometimes the birds are enticed, and if the shooters are lucky each will drop one. There is no greater thrill in marsh hunting than to bag this kind of waterfowl.

Unless ideal weather for ducks arrives near sundown, hunting then is not as productive as in the morning. Waterfowl appear to judge closing time very accurately, so much so that old timers claimed ducks carried watches. At any rate, they move into the marshes immediately after shooting becomes unlawful. I have heard many a hunter swear that, as he was picking up his decoys, the birds flew in so close he could have knocked down his limit with a punt pole.

When marshes were baited with corn, buckwheat, or other grains, birds streamed into the baited waters all day long and quotas could be bagged with little difficulty. This practice was made illegal all over the United States in 1951.

Toward the close of the season, when older and warier ducks predominate, more care must be taken in the placement of decoys and the appearance of the blind. Empty shell cases and scraps of paper must be carefully removed. Hunters should then be mindful that movement is what warns the birds of danger. All the camouflage clothes in the country will not mask a fidgety shooter.

One of the most exacting and exciting forms of waterfowl hunting, done only in private marshes, is jump shooting from a boat. Usually the boat is a punt, with the shooter in the bow and a retriever, preferably a Chesapeake, in the stern with the punter. The punter must be an expert guide who can follow narrow waterways without getting lost. Silence is the key word as the boat slips along one shoreline. The shooter listens intently for any sound that suggests a duck is about to take off. Many species feed or rest on stream edges, and the shooter can be selective, bypassing females and unwanted species. A good retriever that can work its way through dense cattails is necessary, since few of the birds that are shot actually drop in the waterway.

You never know what will happen on a hunt. One mild November day, I was hunting with Ollie Marquardt in a lightweight metal punt, a type new to me. We tied up in a cattail marsh in about three feet of water, and when two mallards headed past we fired our twelve-gauge shotguns from where we sat, both in the same split second. The recoil knocked our combined weight backward, and water poured into the boat.

By the time we rose to our feet, the boat was resting on the bottom and the water was lapping around our knees. We couldn't wade to shore in the deep mud, so we stood and hoped for rescue. It seemed ages before we heard someone in the distance and Ollie called for help. It turned out to be Paul Haupert and Al Glowacki. Paul shouted, "Where the hell's your boat?" and we answered—as we had shot, in perfect unison—"We're standing in it!"

They transferred me to the top of a muskrat house, and the three of them soon bailed out the punt. Then we moved to another location, where Ollie, who had learned his lesson, deposited me in a wooden blind while he shot from the boat. Lady Luck apparently thought we had gone through enough for one day, because we dropped our limit in a very short time.

President Dwight D. Eisenhower hunted in the Cedar Point Marsh in 1954 and 1958, both times as the guest of club member George Humphrey, Secretary of the Treasury. The president preferred jump shooting, with his twenty-gauge Winchester double across his knees and a retriever in the stern of the punt. Superintendent Cornelius Mominee, who acted as his guide, was much impressed by the President's good sportsmanship, good humor, and shooting ability. Corny reported that Ike's twenty-gauge shotgun seldom missed, but not all visitors to the club were like Ike. Another guide told me about a "big shot" he punted out to a blind: "The guy made himself comfortable, took a paperback book out of his pocket, and said, 'You shoot the ducks.'"

October changes radically about the middle of the month. The first half is much like September, with gradually dropping temperatures driving all but a few snakes, turtles, and frogs into hibernation. On warmer days, several of the large, colorful butterflies and dragonflies can still be seen. Phoebes, chimney swifts, tree swallows and, very rarely, rough-winged swallows still search for insects. A few warbler species remain, including thousands of yellow-rumps and an occasional solitary vireo, while water pipits wander over the dikes and bordering fields.

This is the time of sparrows. Wherever the birder wanders there are white-throated, white-crowned, fox, song, Lincoln's, vesper, Savannah, chipping, field, swamp and the earliest of the tree sparrows

—each in its proper habitat. Their more showy relatives often accompany them: towhee, junco, house finch, purple finch, pine siskin, goldfinch and, of course, the resident cardinal. Most of these birds do not vary much in appearance through the year and can be recognized without difficulty. Also common are starlings, red-winged blackbirds, and grackles in huge flocks that include smaller numbers of cowbirds and rusty blackbirds.

The first cold front after the middle of October brings with it biting northwest winds, a spit of snow in the air, and a skim of ice in the marsh. Snow geese from the Hudson Bay area move through in peak numbers for the year, about ten percent of them in the white phase, the rest in blue. They travel both by night and by day, and their sharp, yapping cries may be heard at any hour. Sometimes they pass over so high they cannot be seen without binoculars. In certain years, for some unknown reason, they stay for a short while in large numbers.

Concentrations of diving ducks—especially canvasback, scaup, goldeneyes, ruddies and common mergansers—raft in the open waters of Lake Erie, Maumee Bay, and Sandusky Bay, but few visit the marshes. Horned grebes, double-crested cormorants, and an occasional common loon fish in the lake near the shoreline. Bonaparte's gulls in great numbers replace the flocks of terns along the beaches, while the only numerous shorebirds remaining on the mud flats are killdeers, dunlins, sanderlings, and, rarely, a red phalarope.

Red-tailed and red-shouldered hawks, accompanied by an occasional early rough-legged hawk, ride the wind southward. At this season they do not bunch up together along the Lake Erie shore, but move southwesterly in twos and threes.

November 1 to 30

Chill Winds and a Tired Sun

Much of this month is dark and dismal, and even the most dedicated outdoor lovers find it difficult to be enthusiastic. A blanket of clouds is normal. The sun's arc is low in the southern sky, and its few appearances are uninspiring. Chill northeast winds sweep over the marshes, bearing a sprinkle of snow or sleet, followed by whistling gales from the north and northwest.

The change from autumn to winter seems all too sudden. During the first week of the month, the countryside appears little different from October. Then the landscape, which has resembled an oil painting dominated by reds, yellows, and warm browns, suddenly becomes more like a charcoal drawing. By the close of the second week, all shrubs and native trees except the oaks have dropped their leaves. Cattails, cane, and marsh grasses are gray with age, and in the pools and canals aquatic plants have sunk back into the mud.

Perhaps this is why, on a trip through the wetlands border, a hiker is suddenly impressed by small patches of color—white, red, blue, or purple. This is the season when the lowly sumac reaches its glory, its fruit clumps standing erect like giant crimson candles. Another bright splash is provided by a ragged bunch of red nightshade berries. Apparently birds do not like these berries, because they remain through the winter. Red bittersweet berries, though less common and less spectacular, are rated higher by human standards, and many clumps end up as decorations. One of the more unexpected patches of red is a display of rose hips. Some of these are wild; others, more showy, appear to be escapees from gardens. Hawthorn trees—or thorn apples, as we used to call them—are not very numerous but they are noticeable. Several species can be found, with the color, size, and number of the red fruits depending on the variety. Some are a great attraction for wintering robins, cedar waxwings, and cardinals.

White berries are found almost entirely on rough-leaved dogwood, which is rarely larger than a shrub. It is not very showy, but it provides food for scores of birds, even tree swallows. The white fruits on poison ivy withstand the cold better—but, of course, are less welcome.

Blue and purple fruits appear on a wide range of vegetation. Those on hackberry trees are closer to black. A large, rather showy blue berry on the dikes and marsh borders is useful to the hiker, warning of a mass of green briar, with its savage thorns. The delicate blue of Virginia creeper is probably the most beautiful of all, but it doesn't last. The bright red stems provide a somewhat startling contrast.

The large pokeberry, also called poke, skoke, or pigeonberry, is in a class by itself. Although it is an annual, it grows to six feet or more and produces close to a dozen large clusters of showy purple berries, very conspicuous. Birds love them and devour them as soon as they ripen. However, the red stems that remain are colorful enough to catch the eye.

Finally, there are abundant purple wild grapes. Many along the shore of Lake Erie were destroyed when high water undermined the rim of trees, and huge stone dikes now occupy the space. But grapevines are still numerous along the inner marsh borders. Like wild strawberries, wild grapes are smaller than cultivated varieties but more strongly flavored. Added to an equal amount of domestic grapes, they make a superb jelly. And, of course, the amateur winemaker loves 'em.

At this season the hiker's clothing picks up a variety of burs and sets adrift the tiny parachutes bearing seeds of milkweed and thistle. These weeds are spotlighted for once in their drab existence when the rising sun strikes their frost-laden stems, adorning them with powdered diamonds.

The average date on which the marsh freezes solid is November 23, but even before that date there are many changes in bird life. The number of long-legged waders drops to a few great egrets and great blue herons. Some of the latter may decide to winter over. The many late migrant songbirds drain off until only a few of several species are left. Tree sparrows will remain throughout the colder months and will become the dominant species. They have abandoned the marsh shelter for woodlots and thickets. Birds are now easier to find because there are no leaves to hide them.

Rails and moorhens, except for a few aberrant individuals, have moved southward and with them the majority of pied-billed grebes and coots. The sight of any shorebird is exceptional. The trailing, eddying clouds of starlings, grackles, and red-winged blackbirds reach their peak. They scour the croplands by day and return to the marsh to roost at night. As they drift along, they sometimes alight in a bare tree, which suddenly appears to have sprouted thousands of black leaves.

Except for great flocks of mallards and black ducks and a few pintails, wigeon, and green-winged teal, few species of puddle ducks remain. Canada geese, however, are fairly numerous. Morning and evening, mallards and black ducks feed in the fields, resting during the day on Lake Erie, in open portions of the marshes, or on grain patches planted in the refuges. In earlier years, these groups were small because food was scarce. However, with the advent of corn-picking machines, which leave a good deal of waste grain, their numbers increased to the point where a gathering at sundown of five thousand geese in a cornfield is not unusual. Goose hunting is permitted inside the Ottawa Refuge and in a five-mile zone border-ing the Ottawa Refuge and Magee Marsh boundaries, and all birds bagged must be reported to state officials at Magee. About two thousand geese are shot each season. This figure doesn't seem so large when you learn that it takes 11,000 hunting trips to bag them—that's an average of five and a half trips per goose.

Each year, aerial counts of Canada geese and other waterfowl are made in all the marshes. Officials of the Crane Creek Wildlife Experiment Station at Magee Marsh conduct two counts each month, September through December. From 1985 through 1990, the number of geese seen on one mid-November day in Magee Marsh and the Ottawa Refuge (including the Cedar Point Marsh) averaged 9,706. Obviously, since Canada geese are constantly com-ing and going, the number that actually passes through the marshes is much greater, and the number bagged is a small percentage of the total.

In the cultivated fields beside the marshes, and occasionally along the beaches, other visitors from the far north appear: northern horned larks, snow buntings, and a few Lapland longspurs. Clean farming and fall plowing have reduced their available habitat, but they still follow the path their ancestors laid down when the edges of the marshes were the only prairies in the area. The runways of meadow mice and shrews become more exposed as the weeds are beaten down, attracting rough-legged hawks from the north and swelling the numbers of red-tailed hawks, harriers, and American kestrels, as well as short-eared owls.

In past years, November was the time for upland game bird hunting, first for quail and later for introduced ring-necked pheasants, which loved the dikes and the borders of the marsh. Now quail are gone and pheasants too few to hunt. Since their habitat is largely unchanged and hunting pressure is light or absent, this decline must be attributed to pesticides, herbicides, or granulated fertilizer, which the birds ingest along with gravel.

Sun-loving woodchucks are now sound asleep in their dens, theoretically to awaken on February 2, Groundhog Day. Actually, they will not stir until March. The hiker occasionally comes across an opossum or a skunk, treating the latter with great respect lest it should panic and forget its manners.

Trappers are beginning to capture muskrats, mink, and raccoons. Money obtained from the sale of hides helps pay the expenses of maintaining the marsh. It also helps keep populations in check. Modern marsh construction and stabilized water levels have created ideal habitats for these mammals and their numbers must be controlled. Muskrats are capable of increasing their numbers by eighty percent in one year; if allowed to outgrow their food supply, they would begin to attack each other, destroy their offspring, and suffer serious diseases. A careful count of dens is made in advance of trapping, and yearly quotas are set to prevent over-cropping. Mink can make serious inroads on other mammals, and raccoons are one of the greatest enemies of ground and cavity-nesting birds; harvesting these animals is an important part of marsh management.

December 1 to 31

Paradox: Violence and Peace

Technically, winter arrives on the shortest day of the year, December 22. But nature is not strictly bound by daylength or the calendar, and the condition of the marsh during this month varies greatly from year to year. In some years, sub-zero temperatures have occurred early in December, and blizzards have piled up a foot of snow over the landscape. December 22, 1989 saw a record low of 19 degrees below zero, with temperatures remaining below zero from December 15 to 24. At the opposite extreme, a severe thunderstorm struck on December 26, 1942, and on December 27, 1959 the temperature rose to 60 degrees; on that day dandelions were in bloom, a few chorus frogs sang, an adventurous garter snake appeared, and winter crane flies fluttered in the air.

People who love the outdoors in all weather must come prepared for below-zero temperatures: heavy underwear, insulated trousers and coats, and extra sweaters for emergencies. The best footwear is the heavy insulated rubber shoe, about eight inches high, designed for ice fishermen. Gloves can be a problem, since fingers must be free to focus binoculars. Of course, gloves or mittens can be slipped on at intervals to allow fingers to thaw, but what can you do when your nose freezes, as mine did on a recent December bird count? Luckily, a woman in the group loaned me a loosely knit scarf to wrap around my face and thaw my nose.

The first snowfall is like a great splash of white paint across a charcoal drawing. Trees seem fewer, ponds disappear, and if you are the first hiker to pass, your footprints look like the famous first steps on the moon. If it weren't for the footprints of birds and various mammals, you would feel utterly alone. No doubt it was this abrupt change that caused the ancients to believe that birds bury themselves in the mud of marshes in the fall and emerge when spring unlocks them.

In fact, surprisingly large numbers of living creatures are about —some perhaps peering at you from a sheltered spot. On the Audubon Christmas Bird Census, taken each year at the Ottawa Refuge and nearby areas, the average number of birds seen each year is 26,667. Results of the census vary from year to year according to the weather, the supply of available food, and probably the conditions that prevailed during the late fall migration. The winter population is made up of three groups: permanent residents, birds that normally migrate south but for some reason have remained behind, and visitors from the far north such as Lapland longspurs, snow buntings, tree sparrows, rough-legged hawks and, hopefully, snowy owls. Rarer birds sometimes wander in from other parts of the continent, such as unusual gulls and waterfowl, wheatear, and varied thrush.

Don't expect to find large numbers of birds on every trip. The snow-topped marsh is desolate, and the number of weed seeds available for songbirds is limited. But the hiker who can find a bit of open water in Lake Erie may see great flocks of diving ducks. Before the creation of the Crane Creek Wildlife Experiment Station at Magee Marsh in 1956 and of the Ottawa National Wildlife Refuge in 1961, very few Canada geese or puddle ducks stayed on after the marsh became locked in ice. Now, because they are protected and supplied with plentiful waste grain in the fields, numbers have built up to an average peak of ten thousand Canada geese, nine thousand mallards, and seven thousand black ducks.

Normally, the most numerous perching birds are tree sparrows, starlings, juncos, cardinals, song sparrows, and goldfinches, in that order. In some years there are small flocks of red-winged blackbirds, often accompanied by a few grackles, cowbirds, or rusty blackbirds. Although black-capped chickadees are rare or completely absent during the summer months, there is evidence that they sometimes cross Lake Erie from Ontario and spend the winter on the Ohio shores, along with white-breasted nuthatches and downy woodpeckers.

Some of the most surprising wintering species are warblers: orange-crowned and pine warblers, ovenbird, northern waterthrush, and common yellowthroat. These rarely survive the winter, perhaps because of some disability. I once performed an autopsy on a white-throated sparrow that had died during the winter and found a broken wing that had healed improperly. The bird apparently had been able to fly, but not far enough.

December is one of the best months to get acquainted with large predators, which are more numerous over the marsh and its borders than in any other section of northwestern Ohio. The Audubon Christmas Bird Census for 1974 listed thirty-seven red-tailed hawks, twenty-nine rough-legged hawks, one harrier, two bald eagles, and twenty-seven American kestrels in and near the marshes. No doubt raptors are attracted by the thousands of ducks and geese that assemble in the Ottawa Refuge and the Magee Marsh. Natural mortality and wounded birds from adjacent hunting areas, plus a good supply of meadow voles and other small mammals, provide a steady and easily obtained source of food.

Sunrise at Ottawa National Wildlife Refuge. *Credit: Charles D. Wilson.*

A dike in the Ottawa National Wildlife Refuge. *Credit: Mark Witt.*

Cedar Point Clubhouse, 1922. *Photo courtesy of Tom Mominee.*

Cedar Point Clubhouse, 1970. *Credit: Lou Campbell.*

Erosion near Little Cedar Point, 1962. This entire stretch of beach is now bare sand covered by a huge stone dike. *Credit: Lou Campbell.*

Lake Erie shoreline near the Lamb triangle, 1991. A small remnant of the trees that once lined the entire beach. *Credit: Claire Gavin.*

Duck-hunting blinds are transported along a canal in Magee Marsh.
Credit: Ohio Division of Wildlife.

Ice-fishing shanty. *Credit: Ohio Division of Wildlife.*

Bur Marigolds in Magee Marsh. *Credit: Lou Campbell.*

Bur Marigold. *Credit: Lou Campbell.*

Five-lined skink. *Credit: Larry Hiett.*

Tree Frog. *Credit: Lou Campbell.*

Water Lily. *Credit: Lou Campbell.*

Flowering Rush. This flower was first discovered in the Ohio Lake Erie marshes by the author. *Credit: Lou Campbell.*

Prothonotary Warbler.
Credit: Lou Campbell.

Female Belted Kingfisher. *Credit: Lou Campbell.*

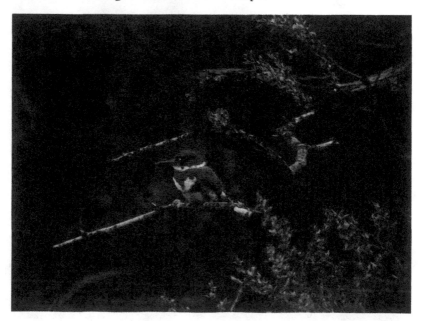

The red-tailed hawk is most common in the open, and some of the juveniles are very tame. The spacing of striped areas on the breast is the best field mark. In most years, rough-legged hawks move down from the north in both color phases, the brown-and-white and the striking black phase. Their habit of hovering to inspect a favorable site is an excellent clue to their identity, since the rough-legged and the osprey are our only large hawks that hover. Harriers fly back and forth over the meadows and the grassy portions of the wetlands, searching for small mammals. This flight pattern, together with the long wings and tail and the white rump, make identification easy. The small, brightly-colored kestrel, with its characteristic head markings, usually prefers utility wires and poles or treetops beside roadways. Like the rough-legged hawk, it hovers while hunting. In the swamp forests, red-shouldered and Cooper's hawks are rare and the northern goshawk is accidental. A pair of bald eagles perched side by side near the top of a tall tree can be seen from far away. Adults display a white head and tail, but juveniles do not completely acquire this plumage until their sixth year.

Nature has conveniently divided the owls into two groups for identification: those with feather tufts called horns, or ears, and those without. Except for the short-eared and snowy owls, all spend the day in swamp forests or dense thickets. Short-ears hunt by day and can be seen coursing erratically over the short-grass marshes and meadows at dawn and dusk. The tufted group includes screech, great horned, and long-eared owls, while the round-heads include barn, snowy, barred, and saw-whet owls. The short-eared owl is an intermediate, with ears that are visible only at short range.

All are easily identified if the observer has an opportunity to examine them, but too often all there is to see is a large bird flying rapidly away. The great horned and snowy owls are the largest, followed by the barn (or monkey-faced) and barred owls. The bass hooting of the great horned owl is a feature of winter evenings. The long-eared owl is almost a replica of the great horned, but about a third smaller, with brighter colors and proportionately longer ears. Screech owls, the smallest of the eared owls, come in gray and red color phases; sometimes the two are found in the same brood. A favorite home for a screech owl in winter is a wood duck nesting box. Snowy owls are unmistakable because of their color and size. A few are seen nearly every year, and occasionally a large flight occurs. In their northern range they seldom see people, and are therefore comparatively tame when they visit northern Ohio.

The barn owl has now become very rare, and barred owls are not found along the marshes at all any more. Gray in color, with dark stripes and brownish black eyes, the barred owl has always reminded me of an old lady with a shawl over her head. Despite its large size, it feeds on small creatures and seldom preys on game birds or mammals. In spring its loud hooting is like no other sound in the wild. It now breeds in two or three areas in the Oak Openings west of Toledo.

Owls are equipped with a novel and efficient method of eating. It is best developed in the medium-sized species like the barn and the long-eared, which live mainly on small mammals. Instead of skinning their prey and laboriously picking the flesh from the small skeletons, they simply swallow it whole and let their stomach acids do the work. After the meat has been dissolved, the bones are wrapped in the undigested fur, and the entire neat pellet is regurgitated. These pellets can be found wherever owls roost for any length of time. By taking them apart and examining the skulls, one can see what the owls have been feeding on.

December offers a great opportunity to explore the marshes and surrounding countryside. Weeds on the dikes are flattened and thickets are more open to travel. After the marshes have frozen, an inquiring hiker can visit places virtually inaccessible at any other season. It's a time to lay out trails for the future and become acquainted with the varying topography of individual marshes. It's also a splendid chance to spot the large nests of crows, hawks, great horned owls, and green-backed herons. Now that the leaves have fallen, the hiker can note which of the woodlots that appeared so dense in summer contain only saplings and which are made up of mature trees.

December marshes have one last gift to offer: silence. The jaded refugee from noise pollution can find a few hours of quiet in the marsh where the only sounds are the ever-present whisper and squeak of the wind through cattails and reeds and its faint moan as it moves in the bordering treetops. Occasionally the silence is broken by the melodious calls of tree sparrows or the staccato pounding of a downy woodpecker.

Silence and the opportunity to look across miles of open space have a healing power that counteracts the inhibitions and muddled concepts fostered by daily routine. It is good to be able to scan the many miles of marshland without seeing artificial structures other than dikes. Even though it is, of course, desirable, civilization is physically and mentally confining. Cities, offices, factories, dwellings—even rooms—imprison the body just as customs and manners constrain the mind.

Viewing the marshes, we can strip ourselves of all these limiting aspects of civilization (except, of course, our warm, comfortable clothes). We can be alone and unconfined. We can enjoy long, deep thoughts unhampered by the interruptions of everyday living and undisturbed by unnatural objects or sounds. Silence and space cleanse the spirit and leave it fresh.

John Burroughs phrased it in a single sentence: "I never found the companion that was so companionable as solitude." Montaigne, three hundred years earlier, was even more emphatic: "In solitude alone I know true freedom."

PART II

The Marshes of
Southwestern Lake Erie

Chapter 1

The Creation and Development
of the Lake Erie Marshes

Like all the landscapes of northern Ohio, the Lake Erie marshes owe their existence to the glaciers of the Ice Age. In at least four separate advances over hundreds of thousands of years, the glaciers scraped soil and gravel from the northern Great Lakes region and carried them off to the south. By the end of the Ice Ages, much of northern Ohio was covered with well over fifty feet of fertile glacial till. As the last glacier melted away to the north, the waters at its retreating edge formed a lake in and around the shallow western end of the Lake Erie basin.

When these waters drained away, they left behind a landscape shaped by flat underlying bedrock and washed smooth by thousands of tons of lake water: an extremely level plain of deep glacial soil, covered in some places with a thin layer of clay and broken only by ancient, abandoned sand and gravel beaches. A strip of this exceptionally flat, rich land along the southwest shore of the lake, regularly flooded by even modest rises in the lake level, became the Lake Erie marshes.

The Gift of the Glacier

During the Pleistocene Epoch (the Ice Ages), which began about a million years ago, a mountain of ice two or three miles high accumulated in central Quebec.[1] The weight of the ice was so great that it depressed the earth's crust by hundreds of feet in much of Canada and the northeastern United States. Under the pressure of this massive weight, the bottom layer of ice softened and bulged all along the glacier's southern edge. Year after year it grew, aided by the earth's cooling climate and its own white, sun-reflecting surface. Over hundreds of thousands of years, the Great Lakes region experienced at least four such ice invasions, each one followed by a period of warmer weather that melted the ice sheet back to Canada.

The last ice advance—called by geologists the Wisconsinan Glacier—began about 50,000 years ago, spreading south and southwestward at an average rate of 100–200 feet a year.[2] When the glacier reached the ancient river valleys that are now the basins of the Great Lakes, it sent into them huge tongues of ice, scouring the softer rocks and dislodging boulders, deepening and widening the valleys into basins. Northern basins, where the ice was thicker, were most deeply scoured. Further south, in the Lake Erie basin, the soft shale at the central and eastern ends was gouged fairly deeply, but the more resistant limestone and dolomite bedrock at the western end caused it to remain shallow. Steep outcrops of this bedrock eventually became the Lake Erie islands.

About nineteen thousand years ago, the glacier extended nearly to the Ohio River. However, the climate had already begun to warm again.[3] The summer sun melted the leading edge of ice faster than winter snows could replenish it, sending the bulk of the glacier into a long, slow, and disorderly "retreat" toward the northeast. Slow only in human terms, that is—in geological time, the five thousand-year journey across Ohio was amazingly fast. The entire glacier was shrinking, lightening the load on the earth's crust, and the land to the northeast slowly began to rebound. Whenever the warming trend was interrupted, as it often was, by a cold spell of a few hundred years, the glacier halted and sometimes readvanced before melting back again; these erratic maneuvers can be traced today in moraines, mounds of earth and gravel debris washed out of the melting ice. One such moraine, stretching all across north central Ohio, is high enough to divide Ohio's waters into north and south-flowing streams (Fig. 1).

When the glacier finally retreated north of the Ohio divide, the mountain of ice became a dam for the water at its melting southern edge. Blocked by the ice from draining away to the north or east, and blocked by the moraine from draining south, water from rain, snow and melted ice collected at the glacier's edge, forming a large lake called Lake Maumee (Fig. 2). This first ancestor of Lake Erie flooded over the broad, shallow western basin all the way to Fort Wayne, Indiana before it found an outlet that led eventually to the Mississippi River. The elevation of any lake's outlet, like the height of the overflow outlet in a bathtub, sets an upper limit on the water level in the lake. Because the western outlet of Lake Maumee was at a much higher elevation than today's outlet to the east over Niag-

Figure 1: The position of the ice sheet just before the formation of the Great Lakes, 14–15,000 years ago. The ice edge across Ohio marks the position of the Fort Wayne Moraine, the watershed dividing north-flowing from south-flowing streams. After Hough, 1958.

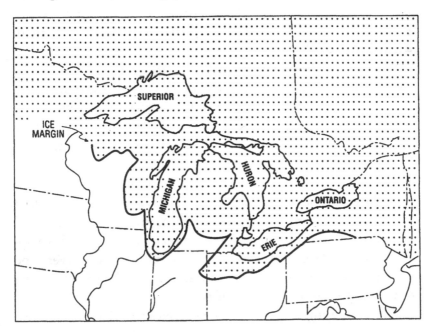

ara Falls, Lake Maumee was over two hundred feet higher than the present Lake Erie.

Gradually, over the next 1,500 years, still retreating and re-advancing, the glacier melted enough to uncover the entire Erie basin. Along the way, new and lower outlets were sometimes uncovered, only to be buried again in the next advance. Each time the outlet changed, the lake level changed. Of course, the lowest outlet of all would have been to the east, where the ice-burdened land still sagged 150 feet lower than it is today. Consequently, when the Erie basin was finally free of ice, its waters were poised to flow eastward, held back only by the ice mass in the Niagara River valley and the Lake Ontario basin (Fig. 3).

Finally, one warm summer's day some 12,500 years ago, the Niagara ice melted just enough to allow a trickle of lake water to escape eastward.[4] This trickle melted and eroded more ice, increasing the flow until quite suddenly the ice dam let go and a tremendous outpouring of pent-up lake waters combined with the melted ice to form a great flood that rushed through the valley and on into the Ontario basin. The flood washed away much of the glacial till that had filled the Niagara Gorge and cut many other channels, most of them now empty. It also completely drained the western Erie basin. Lake Erie became two small bodies of water, filling only the deepest parts of the central and eastern basins, connected by a small stream (Fig. 4).

Most of the drained parts of the basin were soon covered by swamp forest. According to geologist Jane Forsyth, the first swamp forests were probably dominated by black spruce, with some white cedar, willow, poplar, quaking aspen, and fewer elm and red maple. Drier, higher sites probably had some white spruce, fir, and hemlock. Through the centuries, as the climate softened, these species were gradually replaced by hardwoods. Variations in the depth of western Lake Erie, as they appear on today's charts, suggest that the landscape must also have included marshes, wet prairies, and ponds. The Lake Erie Islands and the many reefs would have appeared to an observer as rocky hills on the plain. The rivers and streams that once fed the lake now accelerated their flow, cutting beds into the clay and bedrock of the basin floor.[5]

Meanwhile, the ice continued to recede and the northeastern

land mass continued to rebound, raising the eastern end of the Erie basin. As the outlet rose, the eastern basin filled with water, followed by the central basin and, finally, about 4,000 years ago, the shallow western basin (Fig. 5).[6] Beyond the western shore lay Ohio's lake plain, once the bottom of ancient Lake Maumee and now a flat, clay-covered land rising away from Lake Erie by only two or three feet per mile.

The Creation of the Marshes

Marshes flourish on the western border of Lake Erie because of the poorly drained soils and the extreme flatness of the land. The first Lake Erie marshes were probably created when the slowly rising lake waters finally began to creep into the western basin. The deepening water gradually infiltrated the swamp forests that had covered this basin for eight thousand years, killing the trees at the water's leading edge. The water-killed trees were first replaced by shrubs, but as the lake continued to rise, the shrubs gave way to a succession of increasingly water-tolerant plants: cane and marsh grasses, then cattails and, finally, aquatic plants (Fig. 6). These marsh plants formed a border between the advancing lake and the retreating forest. When the lake finally reached its present boundaries, the marsh border became a permanent but shifting buffer zone occupying the low, flat land between the lake and the huge swamp forest that later became famous as the Black Swamp (see Fig. 10, p. 104).

Advancing lake waters flowed into the mouths of rivers and streams, flooding them to the tops of their valleys for miles inland. Protected from the heavy waves of the open lake, marsh plants invaded the edges of these drowned stream mouths, or estuaries, and their adjacent flooded flatlands. The mouths of the Maumee and Sandusky Rivers became bays that sheltered and nourished masses of aquatic plants, while cattail marshes lined their shores. Even the smaller streams, like the Toussaint River and Turtle Creek, still widen as they approach Lake Erie, though their mouths have narrowed greatly.

* * *

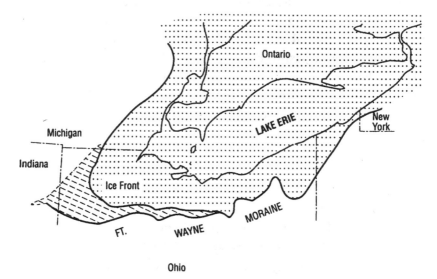

Figure 2: Lake Maumee, first ancestor of Lake Erie, 14,000 years ago. Herdendorf, 1989.

Figure 3: Lake Lundy, 12,400 years ago. Notice that the ice has retreated as far as the Lake Ontario basin. Herdendorf, 1989.

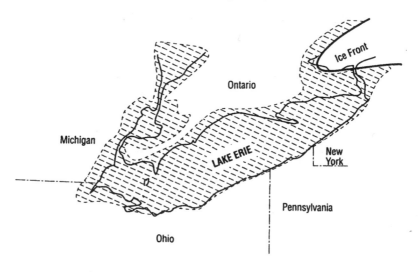

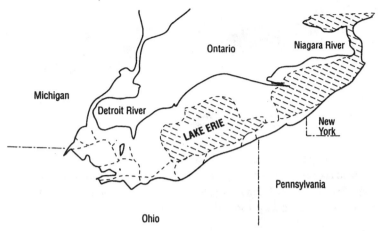

Figure 4: Early Lake Erie, 12,200 years ago. The lake waters have flowed down into the newly uncovered Ontario Basin, leaving water only in the deepest parts of the Erie Basin. At the west end, rivers (-----) are flowing faster, cutting deep beds to reach the remaining lake waters. Herdendorf and Bailey, 1989.

Figure 5: Present Lake Erie, 4,300 years ago. With the weight of the ice removed, the land to the east has rebounded, slowing the flow of water over Niagara Falls and allowing the Erie Basin to fill up to its present level. Herdendorf, 1989.

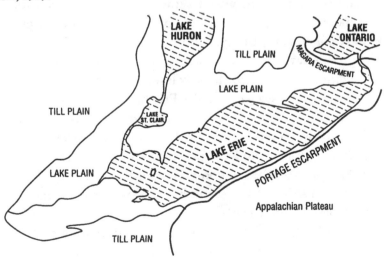

| Swamp Forest | Seasonally Flooded Flat | Wet Meadow | Shallow Marsh | Deep Marsh |

Figure 6: Wetlands vegetation growing from upland swamp forest (left) to an undiked lakeshore (right). Each plant community seeks its own water level. As Lake Erie gradually filled its western basin, this scene moved farther west until the lake reached its present boundary. Adapted from Herdendorf, 1987.

Meanwhile, along the shore of the open lake, drifting sands built up sandbars at the outer edge of the shallows. Many of these sandbars were high enough to trap water behind them in shallow ponds. Wherever the sands were wide and stable enough to support trees, they were invaded by red ash, red maple, box elder, cottonwood, and several species of willow. Behind the tree-lined sandbars marshes formed, protected from the lake (Fig. 7).

Periodically, however, a violent storm would raise Lake Erie's waters above the sand barriers, uprooting and killing both marsh plants and trees. Subsequent wave action would rebuild the sandbars and the vegetation would return—quickly to the marshes, much more slowly to the beaches—only to be destroyed again by the next big storm. This cycle occurred over and over, and continues to this day wherever the shoreline is unprotected.

In addition to the extraordinarily level terrain at its western end, two other factors peculiar to Lake Erie have contributed to these destructive episodes. One factor is the dramatic effect of storm winds on the water level of the lake; the other is the Detroit River, which funnels water into Lake Erie from the upper Great Lakes. A closer look at each phenomenon will help explain why today the southwestern shore of Lake Erie is ringed with twelve and a half miles of massive stone barriers.

Erie is a long, narrow, shallow lake, 240 miles long by only 57 miles at its widest point. Not only is it by far the shallowest of the

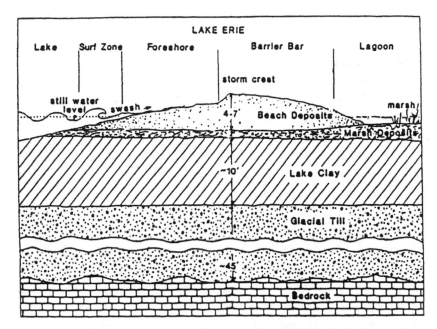

Figure 7: Cross section of a typical Lake Erie barrier bar. Herdendorf, 1987.

Great Lakes, averaging only 25–35 feet in the western basin, but its bottom is generally smooth and flat, with few ridges to obstruct wave action. As a result, high winds on the lake are readily translated into high waves. Winds on Lake Erie are given still greater force by the lake's southwest/northeast orientation, since the strongest storm winds that hit the lake are precisely those that approach from these directions. Such storms can whip up fifteen-foot waves, which break offshore and reform into shoreline waves several feet high.[7] A strong wind out of the northeast can raise the water level along the southwestern shore by five or six feet over and above the height of the waves. And these waters are breaking on a shore so flat that a two-foot rise in lake level could flood for a mile inland if there were no barriers.

Although northeast winds create the worst floods, even persistent southwest winds can raise the water level of western Lake Erie, because of a phenomenon known as a seiche (pronounced "saysh"). A seiche begins with sustained storm winds that raise the water level at one end of the lake. When the wind dies, the water level drops at that end and rises at the other, oscillating from end to end like the slosh of water in a bathtub. The shallowness, smooth bottom, length, and orientation of Lake Erie all combine to create the worst seiches in all of the Great Lakes.[8] For example, at the peak of a northeast gale, 10:00 A.M. November 14, 1972, the elevation of the west end of Lake Erie was 576 feet above sea level; by 7:00 P.M. the wind had died and the water had fallen to 572 feet above sea level—a drop of four feet in only nine hours.

The second major influence on Lake Erie levels is exerted by long-term changes in the Great Lakes watershed. Ninety percent of Lake Erie's water comes from Lakes Superior, Michigan, and Huron via the Detroit River. This funneling of water from the upper Great Lakes means that even small changes in their water levels can create big changes in Lake Erie. Water levels in the three bigger lakes have always risen or fallen in response to extended periods of rainfall or drought. For example, in the eight-year period from 1965 to 1973, above-normal precipitation in the Great Lakes watershed led to increased runoff, while below-normal temperatures decreased evaporation.[9] The result was higher water levels in all the Great Lakes, and record highs in Lake Erie.

A more disturbing cause of rising waters in recent years is the population boom in the Great Lakes watershed, which has replaced more and more trees, pasture, beaches, and wetlands with bare, clean-plowed fields, residential areas, shopping centers, and four-lane highways. An immediate result is faster runoff and higher peaks in lake levels. In addition, loads of silt have reduced the carrying capacity of the basins, especially Lake St. Clair and the western end of Lake Erie.

Perhaps it is not a coincidence that changes in Lake Erie levels are more pronounced than they were. In the sixty-five years from 1860 to 1925, average yearly water levels ranged from 570 to 571.5 feet above sea level, a difference of 1.5 feet (Fig. 8). In the fifty-three years from 1934 to 1987, average yearly levels ranged from 568 to 574 feet—a difference of six feet. Of course, even at moderate lake levels, a severe storm can cause floods: in 1929 and 1969, when average levels were about 571 feet, northeasters raised waves that broke through barriers and flooded thousands of acres of marsh and farmland.

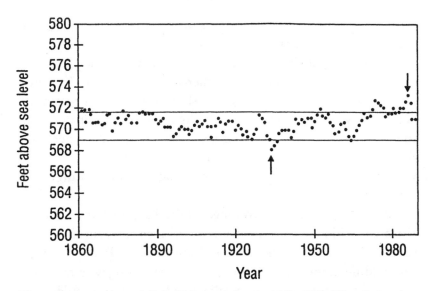

Figure 8: Average yearly Lake Erie water levels, 1860–1989. The all-time low was in 1934, the all-time high in 1986.

The greatest potential for destruction arises when an overall increase in average lake level combines with seasonal high waters and a severe storm. This is exactly what happened on April 9, 1973, when forty-five mile-an-hour northeast winds raised the waters of western Lake Erie to 578 feet above sea level, or four feet above the record high yearly average, *plus* high waves.[10]

Without huge stone barriers to protect the lakeshore, high water and severe storms would long ago have destroyed all the existing marshes and much of the farmland near the lake. In past ages, new marshes would have been created further inland, but human beings have other plans for those areas and are fighting to confine Lake Erie to its present boundaries.

Throughout the years before the barriers were installed, when every shift in lake level altered the depth of water in the marshes, any long-term change in the water supply reverberated throughout the entire marsh ecosystem. Though marsh plants are well adapted to short-term fluctuations in water level, each plant community thrives at a particular *average* water level. As a result, even moderate long-term changes can transform whole sections of a marsh. For instance, when the water level increases, cattails shift inland to stay in the belt of shallow water, and are replaced by aquatic plants like water lily. This is what happened when the water level in Lake Erie rose at the end of the Ice Ages, pushing the marshes westward. When the water level falls, the cattails again seek their optimum conditions, this time by moving lakeward, and their former position is taken by bluejoint grass and cane. This process is easy to visualize by referring back to Figure 6 and imagining a higher or lower waterline.

Such changes were not confined to the Lake Erie marshes, but also occurred in the marginal wetlands of such sluggish streams as Crane and Turtle Creeks and the Toussaint and Portage Rivers. The wildlife in the marshes must have varied along with the vegetation. For example, cane and solid stands of cattail provide such limited food that periods of low water must have greatly diminished marsh wildlife.

Hints of past transformations can be found in some of the first descriptions of the western Lake Erie shore. Early explorers mention vast expanses of cane—which they called "prairie"—along the western Lake Erie shore. John D. Riddell, writing in 1837, mentions the "Grand Maumee Prairie" beside Lake Erie "from the Maumee River to the headwaters of the Portage."[11] Still earlier, in 1815, Samuel Brown described an area "ninety miles long and two to ten wide, extending from the mouth of the Portage" nearly to Detroit. There they found grass "higher than our heads and as thick as a mat confined together by a species of pea vine." Near the mouth of the Toussaint River, in what is now the Toussaint Marsh, the prairie grass was "about seven feet high and so thick that it would easily sustain one's hat—in some places a cat could have walked on its surface."[12] These observations suggest that some time before 1800, an unusually large area of swamp forest was destroyed by high water, after which a period of low water encouraged the growth of cattails and cane in the cleared spaces. Modern marshes are somewhat cut off from this cycle by the dikes that separate them from Lake Erie. However, exceptionally high or low lake levels can create dramatic changes even in diked marshes: during the dustbowl years of 1934 to 1936, the water level in the Cedar

Point Marsh fell very low, and cane replaced cattails and grass until it occupied the entire inner third of the marsh. In later years, when the water returned, the cattails returned. The high waters of 1973–74, which breached dikes all along the lakeshore, eliminated at least eighty percent of the cattails in the Cedar Point Marsh, and most of the marsh became a huge, shallow pond. Figure 9 illustrates the effect of high Lake Erie levels on the Toussaint Marsh.

Through the ages, as we have seen, the Lake Erie marshes have been bordered on the landward side by a stable hardwood swamp forest of oak, hickory, maple, chestnut, and black walnut. Most geologists consider this forest part of the Black Swamp (Fig. 10), which extended far to the southwest, interrupted only by limestone ridges and the ancient sandy beach known as the Oak Openings.* In these days of waving cornfields, it is difficult to imagine the immensity of the Ohio forest, and especially of the Black Swamp, before logging began. Apart from the Oak Openings, the only breaks of any importance in the dense woodlands were made by water-filled depressions holding wet prairies or stands of buttonbush. Between the swamp forest and the marshes was nearly always an impenetrable fringe of fast-growing shrubs: willows, dogwood, sumac, and saplings of ash and box elder.

The massive stumps and the occasional log section that can still be found in the few remaining patches of swamp forest reveal a long growing period. The uniform width of the rings suggests that comparatively little change in overall growing conditions has occurred in more than a century—a sharp contrast to the ever-changing marshes. The forest growth is unaffected by the pools of surface water that often remain from late winter through spring, although prolonged high water would kill the trees within five years.

---

*The Oak Openings, west of Toledo, is a broad ridge of fine yellow sand extending from the Maumee River to Detroit. The sand was deposited by several of Lake Erie's Ice Age predecessors, each of which was higher than the present lake. The Openings, named for its sparse vegetation and easy travel compared to the horrors of the Black Swamp, is of great historical and ecological interest, though outside the purview of this book.

Figure 9: Three views of the Toussaint Marsh, from aerial photographs taken in 1950, 1970, and 1977. These three views illustrate the effect of high lake levels on marshes that are not heavily diked.

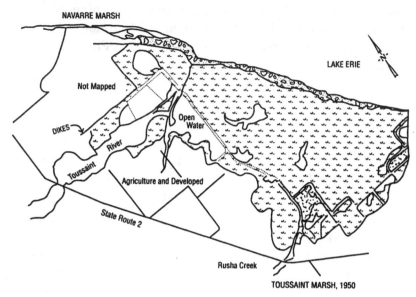

9A: In 1950, cattails and other shallow-marsh plants made up 73% of the marsh, open water 10%. A fringe of hardwood trees bordered the marsh on the lakeward side.

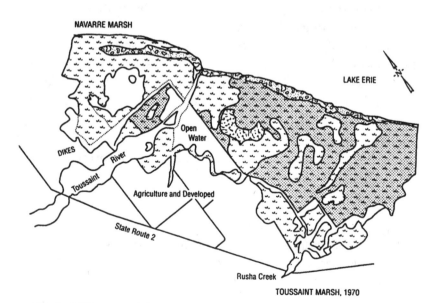

9B: In 1970, most of the cattails were water-killed. Floating and submerged aquatic plants appeared, but still made up only 3.5% of the marsh.

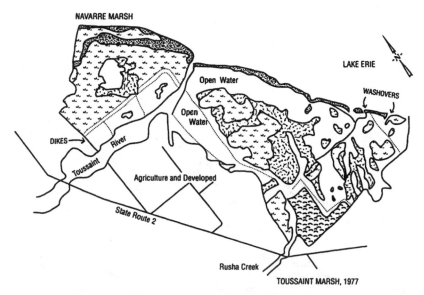

NAVARRE MARSH

LAKE ERIE

Open Water

WASHOVERS

Open Water

DIKES →

River

Toussaint

Agriculture and Developed

State Route 2

Rusha Creek

TOUSSAINT MARSH, 1977

9C: In 1977, open water covered 40% of the marsh, floating and submerged aquatic plants another 20%, and cattails and other emergent plants only 16.5%, or one quarter of their 1950 acreage. The hardwoods along the lake gave way to more water-tolerant shrubs.

Cattails

Water Killed Cattails

Meadow

Aquatic Plants

Hardwoods

Shrub

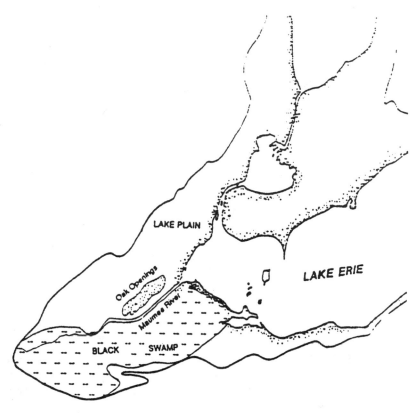

Figure 10: The infamous Black Swamp extended the length of Ohio's Lake Plain. *Adapted from Forsyth, 1960 and Herdendorf, 1989.*

The immense size of the original trees, notably the white oaks, made them very attractive to early lumbermen for ship timbers and dock piling. Less valuable trees were burned for charcoal or laid in rows to make corduroy roads. Although by 1875 intense logging and burning had entirely cleared the Black Swamp, remnants of swamp forest may still be found in the Ottawa National Wildlife Refuge, Magee Marsh near Turtle Creek, Navarre Marsh, and Toussaint Marsh. These woodlots also contain a number of cottonwoods, which moved in when the dense swamp forest was cleared. There have been no evergreens since the few red cedars on the outer beaches were destroyed by the high water of the 1970s. However, cedars must once have been plentiful, judging by the names of the two Cedar Points, Cedar Creek, and Cedar Island, now Turtle Island.

Settling the Marsh Borders

The marshes and swamp forests, with their standing water, clinging mud, mosquitoes, and malaria, were clearly not human habitat. The Indians of the area—first the Eries, then the Miamis, and finally the Ottawas—established their communities on higher ground, visiting the marshes to harvest wild rice and various tubers, to trap and spear fish, and to shoot or snare waterfowl. Streams, sand beaches, and cane borders were the principal highways. An Indian trail in the cane border extended from the present site of Toledo to what is now Port Clinton.

Like the Indians, the first white settlers avoided the immediate vicinity of the marshes, except for French trappers engaging in the fur trade. The portions of Lucas and Ottawa counties adjoining the wetlands were the last lands in northwest Ohio to be cultivated. The area's three townships—Jerusalem, Erie, and Carroll—were established in the mid-1800s, but their population growth was slow. Perhaps as a result, the region has been largely ignored by historians; for example, in his otherwise excellent 416-page book, *Lake Erie,* Harlan Hatcher allots to it five lines on one page and two on another.[13]

The largest and most accessible early route through the marshes was the Toussaint River. According to legend, it was originally named "Tous Saint," in honor of All Saints Day, by French missionaries who entered the river in October of 1725.[14] Situated about halfway between the Portage and Maumee Rivers, the Toussaint outlet provided a natural safe harbor for canoes and other boats journeying along the lakeshore. Without a doubt, it was the opening wedge into the vast marshlands.

In addition to its strategic location, the land on both sides of the Toussaint near its mouth was relatively well-drained and accessible. It was first settled by the Erie Indians, who left behind a burial ground, and then by the French. Early maps show two communities on the east side of the river: the first, Bois Blanc, was founded about 1814. The term "bois blanc," or "white wood" was commonly used by French explorers to designate stands of cottonwood. Another such stand, at the mouth of the Detroit River, gave its name to the amusement park, Bob-lo Island.[15]

For some reason Bois Blanc did not survive and was absorbed by a second community, Locust Point, located further south along the river. From 1835 to 1837, a gang of thieves hid in the nearby marsh, using it as a base for raids on the infant community and the surrounding countryside until they were stopped by organized opposition. The area became more accessible about 1878, when a bridge was constructed at the point where Route 2 crosses the river.[16] Since a settlement a few miles west at the mouth of Turtle Creek was also named Locust Point, residents of the Locust Point on the Toussaint River sensibly changed the village name to Toussaint; they became known as "Tousangers."

In 1834, a dispute arose when Michigan insisted that its southern boundary was farther south than Ohio was prepared to admit. If this boundary had been adopted, nearly all of Toledo would now be in Michigan, along with Cedar Point Marsh, half of Ottawa National Wildlife Refuge, and a fraction of Magee Marsh. The arguments flew back and forth for months, and at times it seemed that Michigan was sure to win its case. However, the parties finally reached a settlement in 1836, and Michigan accepted the Upper Peninsula in exchange for giving up its claim to the disputed lands.

One of the region's first major economic enterprises began in the early 1860s, when Eber Brock Ward purchased nearly 8,500 acres of swamp forest hardwoods along what is now State Route 2. Ward built a shipyard and a sawmill near the present site of Bono.[17] In order to ship his logs, he filled in meandering Cedar Creek where it crossed Route 2 and diverted its water into a canal extending over a mile northeast to Lake Erie. Sixty to 150 feet wide and fifteen feet deep, Ward's Canal was dug by hundreds of men using shovels, augmented with horse-drawn single slip and double wheel scrapers.[18] A smaller channel, Cooley Canal, ran from the end of Cedar Creek north to Corduroy Road, and was later extended to Lake Erie. Today, this canal is known as the Reno Side Cut Ditch.

Other lumbermen followed Ward, and at one time the region kept seven sawmills busy. The village of Shepardsville was founded in 1870 to provide homes for sawmill employees, but when its residents later applied for a post office, they were told that Ohio already had a town called Shepardsville. By popular consent, the village name was changed to Bono in honor of a popular Cherokee resident, Frank Bunno. Thirty years earlier, when a large group of Ottawas and Cherokees living just to the east had been removed to Walpole Island, Bunno had refused to leave. Later, during a flu epidemic, his knowledge of herbal medicines and his generosity were credited with saving the lives of many of his neighbors. The town's population probably peaked in the late 1800s; in the 1969 census it was listed at 350.

Clearing the region's forests opened up its rich wet soils, ideal for growing onions; by 1870, Ward's now-abandoned sawmill had become a storage bin for harvested onions. However, the soil remained that of a swamp forest: water-saturated, dotted with shallow pools, and crossed by few natural waterways. Whereas Ward had dug his great canal for transportation, farmers now dug canals and ditches through their fields for better drainage. Virtually all the early roads were bordered with deep ditches, which provided additional drainage as well as earth for the roads.

All of this clearing, draining, and cultivation affected local waters in a way that has since become familiar: the springs and streams that had once run drinkably clear either disappeared or choked on runoff from ploughed fields or eroded banks. Maumee and Sandusky Bays, once lined with bottom-rooted plants—including several thousand acres of wild rice[19]—lost them all as needed sunlight failed to penetrate the muddy water. Only the marsh ponds, which were spared the strong waves that kept bottom sediments stirred up in the bays, retained their full variety of aquatic plants.[20]

Even today, Lake Erie is one of the most important fresh water fisheries in the world. In 1988, for example, this "walleye capital of the United States" produced a record twenty million pounds of walleye, all caught by sport fishermen at a limit of six per day. By at least the late nineteenth century, commercial fishing had become an important part of the local economy. Fishermen maintained operating stations on the outer beach of the Cedar Point Marsh and the mouth of Ward's Canal, taking advantage of the excellent catches in Lake Erie off Jerusalem and Carroll Townships. At the mouth of Turtle Creek, Grow's Dock could be reached either by water or overland along the sand ridge from Sand Beach. Another

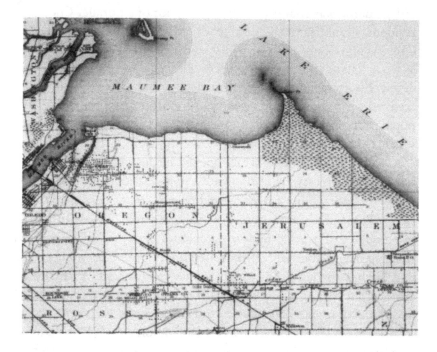

outpost, established about 1865 at the mouth of Crane Creek, was run by James Byers for forty-seven years, from 1888 to 1935.[21] Some of the old foundations of the net sheds and other buildings can still be seen on what is now the property of the Ottawa National Wildlife Refuge.

At the turn of the century, on Lucas and Ottawa County maps,[22] the shoreline is shown as one continuous marsh from Little Cedar Point to Port Clinton. However, the U.S. Interior Department topographical map from 1900 shows a few tongues of better-drained land on the lakeshore east of Turtle Creek, some of which were already settled; for instance, the future sites of the Erie Industrial Park and Camp Perry, northwest of Port Clinton, had a combined total of twenty-five private dwellings.

There were other early settlements that do not appear on these county maps. In 1860, at the junction of Cedar Creek and what is now Route 2, Eber Brock Ward founded a community he named New Jerusalem; it failed to take hold and was soon dropped from official maps. Another small settlement, which persists to this day, is the nameless community established at the intersection of Cedar Point and Decant Roads by descendants of the first French settlers.[23]

After many marshlands were drained—a phenomenon described in the next section—the period from about 1918 to 1925 saw a great expansion of housing developments, from Port Clinton west to Maumee Bay and on up the Michigan shore to Monroe. Early settlements had been laid out on high ground, but now homes near the water became very popular, no doubt inspired by the easy and inexpensive transportation afforded by the mass-produced automobile.

To meet the growing demand for summer homes, several subdivisions were established on drained marshlands between Cooley and Ward's Canals: Reno Gardens in 1918, Reno-by-the-Lake in 1921–23, and Howard Farms in 1921. Together, these subdivisions formed a community larger than Bono, and even supported a golf course from 1921 to 1945. Four other subdivisions were laid out on

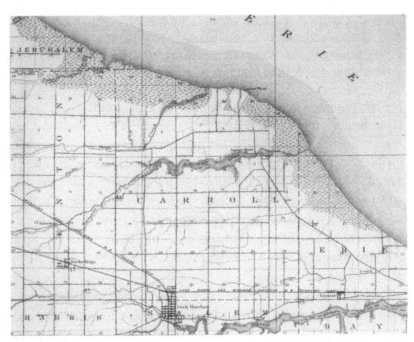

U.S. Geological Survey maps of the Lake Erie marshes, 1900. The high ground at the later sites of Camp Perry and the Erie Industrial Park appears, unmarked, at the southeast corner of the shoreline. Notice how Turtle Creek and the Toussaint River widen as they approach Lake Erie. Shepardsville has changed its post office name to Bono.

Creation and Development of the Marshes   109

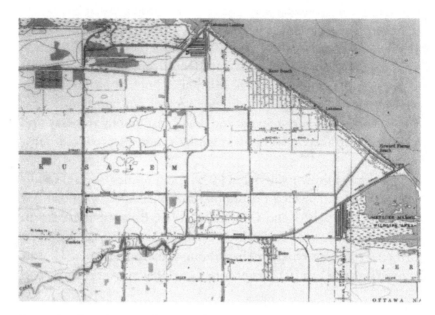

Reno Beach and Howard Farms, former marshland drained for farms and homes shortly after 1900. Note where Cedar Creek (*lower left*) has been diverted from its normal northeasterly course to flow into Ward's Canal (*right*); the low-lying former bed of Cedar Creek bisects the drained land, increasing the flood hazard. *U.S. Geological Survey, 1967.*

higher ground east of Turtle Creek: Locust Point, Sand Beach, Long Beach, and Willow Beach.

Camp Perry itself was established by the State of Ohio in 1906 for National Guard training. Because of its great expanse, it was in demand for shooting club tournaments. Early in 1918, the U.S. government took over the property, adding three hundred acres to its western border and installing a weapons testing section. Nearly every year since, the sound of big guns has been heard for miles, echoing strangely over the peaceful marshes on both sides of the proving ground.

Today, several large marinas are situated beside streams or canals between Toledo and Port Clinton. The newest addition to the lakeshore is Green Cove, a large modern development on the western border of Sand Beach Marsh. As of 1989 it held forty two-story condominiums and a canal, plus a mooring basin for residents' boats. The most prominent addition is the Davis-Besse Nuclear Power Plant, which began operating in 1977. The huge, 493-foot cooling tower, a monument to human efforts to meet energy demands while preserving the environment, can be seen from every Lake Erie marsh. Here one of the newest human technologies rubs elbows with the oldest wildlife community in the middle west.

Curbing the Marshes

By the late 1800s, the clearing and draining of the Black Swamp had made access to the Lake Erie marshes much easier. Hunters had already begun to acquire some of the better wetlands, and other marshes were bought for future farms or housing developments. These low-lying marshlands, with water levels that rose and fell with Lake Erie, presented a special problem for all new property owners: *They had to control the lake.* While farmers and developers intended to drain the marshes completely, hunters and trappers aimed at controlling water levels, hoping to produce a habitat that would increase the muskrat population and entice more waterfowl to breed or visit. Whatever the project, all required protection against Lake Erie.

These diverse plans resulted in a remarkable burst of construction and rearrangement of waterways along the lake from Toledo to Port Clinton. Between 1890 and 1910, clams and draglines, first steam and then gas-powered, were busy the full length of the marsh district, digging channels parallel to the lake and piling up the earth to hold back its waters. (The combination of a channel and its embankment in the marshes is known as a dike, although very often the term refers to the embankment alone.) Some equipment was mounted on barges, which inched forward as the channels were dug; other equipment was operated from the shore. The highest dikes were built on the low sand ridge along the shoreline between Little Cedar Point and Turtle Creek, which was less prominent than the ridge further east and more vulnerable to flooding.

The natural waterways through the marshes, themselves subject to seasonal floods, received their share of attention. Crane Creek, which meandered with many curves toward Lake Erie, was straightened from Stange Road to its mouth at the lake. The lower stretches of the Toussaint River and Turtle and Lacarpe Creeks were confined between high banks. Lacarpe Creek, which flows into the Darby Marsh, had two outlets into Lake Erie in the late 1800s, but was cut back to one by 1900. Double rows of wood or metal piling were driven at the mouths of all important streams, reaching far out into the lake to prevent shifting sands from blocking channels.

In 1902, Henry Reno and George Howard purchased nine hundred acres of Cedar Point Club property east of Ward's Canal. Most of the land consisted of shallow marsh, lying along the lakeshore behind low sand beaches and cut by even lower ground, the former bed of Cedar Creek. Despite its vulnerability to flooding, Reno and Howard transformed the marshland into onion farms. To drain the property and form its western boundary, Cooley Canal was extended from Corduroy Road north to the lake. A dike was built along the three-mile shoreline between the two canals. Later, this land became Reno Beach, Howard Farms, and a smaller development known as Lakeland. East of Ward's Canal, other drained marshes became Metzger and Pintail farms, both shown in a 1924 county atlas as the property of the National Land Company. The decision to drain these low, flood-prone areas was a serious mistake, as we shall see. Scattered marshlands inside the relatively high sand ridge between Turtle Creek and Sand Beach were also drained, with better success.

Meanwhile, sportsmen opened their undeveloped marshes to boat travel by dredging channels through them. The channels eliminated one of the great handicaps of early hunters and trappers, who had found it almost impossible to reach the marsh interiors. Some owners planted willows and encouraged other vegetation to grow on the channel banks for erosion control, while others kept the dikes cleared for travel, sometimes shoring them up with crushed stone. Most of the large marshes now have stone roads on many of their embankments, which continue to provide access to marsh interiors long after the canals have become choked with silt.

The dikes winding in and around the marshes, together with the large pumps installed to maintain water levels, added a definite modern touch to the landscape. However, they did not detract much from their primitive surroundings, and did bring definite improvements: they stabilized water levels at the optimum level for muskrats, nesting waterbirds, frogs, reptiles, and insects; they created more of the "edge habitats" preferred by nesting waterfowl; and they allowed marsh owners to manipulate water levels in order to encourage the growth of waterfowl food plants. Several species of duck began nesting in the Lake Erie marshes only after the dikes were built. The dry soil of the banks added an additional habitat to the marsh complex, one that attracted ducks in the breeding season, game and songbirds all year round, a variety of mammals and snakes, and turtles looking for a place to lay their eggs.

However, from the standpoint of waterfowl and fish production, the dikes brought some less desirable effects: they invited raccoons, skunks, opossums, and snapping turtles, all of which destroy duck nests or ducklings, and they completed the disappearance of wild rice, good duck food that requires clear, moving water. The outer embankments seal off the marshes from Lake Erie fishes that once entered to spawn: northern pike, largemouth bass, crappies, bluegills, pumpkinseeds, catfish, bullheads and forage minnows. Ultimately, the enclosed marsh ponds and canals contained mainly carp, goldfish, and in some areas black bullheads, bowfins, and mudminnows.

The inland canals proved to have a useful life of twenty-five to thirty-five years. Bank erosion and the accumulation of decaying vegetation gradually filled them up, leaving two or three feet of water overlying deep, soft mud. Normally the bed of the marshes is shallow enough to invite wading, but any hiker or hunter so tempted must beware of these old stream channels, ditches, and canals. One day in 1953, in the Cedar Point Marsh, I pushed a twelve-foot punt pole full length into the muck without reaching hardpan.

As duck hunters became more numerous, they sometimes bought farms or other well-drained lands and transformed them into marshes by pumping in lake water. Some of these were former marshes that had been drained in the past, others were borderlands that had never been under water. In 1934, when the average level of Lake Erie was eighteen inches below the marshes, all the marsh owners were forced to pump in water from the lake, and many pumps operated day and night. A few of the owners of large marshes found pumping so expensive that they allowed part of their holdings to dry up. Since the late 1930s, of course, high water has been the problem, forcing managers to pump water out of the marshes into the lake.

Even when lake levels are high, pumping is usually necessary in spring and early summer to keep the vegetation in enclosed wetlands supplied with water. Although this is the season when the lake is at its annual peak, growing cattails and other plants utilize a surprising amount of water. Even after the plants reach full growth, their respiration releases a great deal of moisture, which must be replenished. From the beginning, however, the greatest expense and the biggest headache has always been the difficulty of keeping Lake Erie waters in check.

Those first farmers, real estate agents, and cottage owners, as well as duck hunters, soon discovered that the lake is not easily harnessed, as they watched their earthen dikes steadily eaten away by wave action. Accustomed to the generally stable lake levels of the late 1800s, owners had not anticipated Lake Erie's relentless efforts to reclaim the land. During high water periods, northeast storms poured water over the outer embankments. Reno Beach and Howard Farms were flooded in 1929, 1943, 1952, and 1972; Teachout Road was flooded in 1938. Over the years, millions of dollars in state and federal funds were spent rebuilding and enlarging the protective dikes, and the effort continues to this day. Metzger and Pintail Farms, first damaged in 1910, were inundated in 1929 and never reclaimed; they are now owned by the State of Ohio and the U.S. Fish and Wildlife Service, respectively.

The fate of Pintail and Metzger was shared by most of the Lake Erie marshes. Although in early 1951 private gun clubs owned all thirty thousand acres of undeveloped marshland between Toledo and Sandusky, by 1966 the situation had completely reversed: all the largest marshes from the Maumee River to the Toussaint were either owned or managed by the state or federal government. At least part of the reason was the immense cost and effort of rebuilding embankments, re-dredging canals, pumping water in, pumping water out. Most private owners simply couldn't cope with their ever-increasing expenses.

Beginning in the late 1960s, the remaining private owners and the government agencies began hauling massive chunks of limestone from nearby quarries to shore up fragile earthen embankments. But in spite of all efforts at control, the severe storms of November, 1972 and April, 1973 flooded many marshes, farmlands, and homesites between Oregon and Port Clinton. The 1973 storm cut through virtually all the dikes along Lake Erie and its tributaries except most of those that had been rebuilt after the November blow. Fortunately, in most places the floods didn't last long, since the channels cut by storm waves through dikes and sand ridges not only let the lake waters in but let them out again. However, Reno Beach and Howard Farms, some farms near Turtle Creek, and the roadside park at the junction of Routes 2 and 19 remained flooded until late summer. The Toussaint Marsh was still flooded in 1977 (see Fig. 7, p. 97). Total damage and rehabilitation amounted to over a billion dollars.

Not all of the loss was calculable in dollars. The rushing flood waters brought disaster to marsh wildlife. Woodchucks and skunks drowned in their dens, while many hundreds of voles, white-footed mice, shrews, and other small mammals were killed by the water covering fields and thickets. Searching for higher ground, muskrats, rabbits, skunks, and raccoons congregated on roadways, where automobiles and trucks were unable to avoid them. When snakes, frogs, and turtles awakened from hibernation, they, too, sought out higher ground. Naturally deliberate in their movements, many were crushed on the highways.

The dikes undoubtedly saved thousands of lives, especially the broad, substantial barriers in the Ottawa and Cedar Point National Wildlife Refuges and Magee Marsh. In many cases, dikes also protected waterfowl nesting habitats within the marshes. Fortunately, waterfowl production was affected less seriously than it might have been, since the prolonged wet weather had already increased breeding failures.[24]

In response to the devastating storms of 1972 and 1973, ever more massive dikes have been built along the lakeshore, and so far, at least, they are holding. The high water of 1980, which actually exceeded the 1973 high by an eighth of an inch, caused no flooding at all because it could not breach the new dikes. After seventy years of effort, marsh managers seem finally to have succeeded in eradicating the lake's influence over the marshes. The old cycle of destruction and renewal by Lake Erie is gone forever, broken by the barriers that keep the marshes from disappearing altogether.

It seems ironic that in order to save the most primitive places remaining in Ohio, government agencies and private owners have been forced to spend huge amounts of money on modern and sophisticated equipment. However, nothing less would do the job, and it is clear that the lake must be harnessed somewhere—if not along the present lakeshore, then much further inland. Metzger Marsh and the east end of the Cedar Point Marsh illustrate what would happen if the outer dikes had not been built; in these undiked areas, Lake Erie has simply taken over.

Chapter 2

Who Has Owned the Marshes?

In Chapter 1, we saw that the marshlands along Lake Erie arose as a sensitive buffer zone, poised between the essentially changeless habitats of open water on the one side and swamp forest on the other. As an ecosystem, the marshy zone endured over the centuries, even as its various habitats shifted—sometimes abruptly—in response to Lake Erie's changing levels. West of Turtle Creek, the extraordinarily flat land and relatively low sand ridges made the marshes especially vulnerable to flooding.

White settlement began a new cycle of change in these low-lying marshlands: drained for farms or subdivisions, they were repeatedly flooded by rising lake waters, and in many cases allowed to revert to marsh. In 1900, virtually the entire shoreline between Toledo and Port Clinton was marshland, all in private ownership. By about 1905, most of the low-lying marshes had been drained and confined by earthen dikes. Since 1970, the southwestern Lake Erie shore is again lined with marshes, now mostly owned or managed by federal or state agencies dedicated to preserving them. Each of the larger marshes has its own story and its own unique character, forged from a series of changes wrought by Lake Erie, by white settlement, and more recently by the struggle for preservation.

Cedar Point National Wildlife Refuge

The Cedar Point National Wildlife Refuge covers the 2,445-acre, roughly triangular peninsula that separates Lake Erie from Maumee Bay. The marsh extends from the sandbar known as Little Cedar Point eastward about three and a half miles to Cooley Canal and south to the Williams Ditch. Most of the Cedar Point Marsh has never been drained, which makes it the largest single undisturbed marsh on the southwestern Lake Erie shore; its center is undoubtedly one of the most primitive areas in northwestern Ohio. Most of the inland border of the marsh was originally wooded, but the portion west of Yondota Road was cleared beginning in 1961. Off the tip of the sandbar a shoal extends to Turtle Island, interrupted only by the Toledo Ship Channel. Shown as "Cedar Point" on early maps, this tip has been known since 1954 as "Little Cedar Point" to distinguish it from the larger sandbar near Sandusky.

The Point must have been used by many early explorers as a landmark, but the first to mention it was Antoine de la Mothe Cadillac, the founder of Detroit. In 1680, under orders from Count de Frontenac, he traveled up the Maumee River from Lake Erie and built Fort Miami where Fort Wayne, Indiana now stands. The next to record (Little) Cedar Point as a navigation aid was Céleron de Bienville, who in 1749 entered Maumee Bay from the river on his way to Detroit. His fleet of twenty-three canoes and pirogues had to be dragged laboriously over the Maumee River rapids.[1] In 1754, Chevalier Chaussegros de Léry noted the point on an exploratory trip from Erie, Pennsylvania, to Detroit.

The Cedar Point Marsh occurs in the memoirs of James Smith, a white man adopted as a child by the Mohawk Indians. Under the date of March, 1757, he writes, "[We] put in at the mouth of the Miami of the Lake at Cedar Point where we remained several days and killed a number of turkeys, ducks, geese and swans." Returning in November of the same year, he and his companions hunted for several days, bagging thirty deer and numerous wildfowl, especially geese.[2]

Little Cedar Point again figured in history during the War of 1812. On January 23, 1813, a band of refugees from the Raisin River Massacre near Monroe, Michigan crossed the Lake Erie ice to the Point. They were led by Joseph Mominee and accompanied by Robert, James, and Peter Navarre, who had made their escape disguised as Indians. Peter Navarre went on to become locally famous as "Peter the Scout," and Joseph later gave his name to Momineetown. Many descendants of both families still live in Toledo and Oregon.[3]

Later, about 1880, "Ol' Joe" Chevalier operated a tavern near the Point that was a favorite with trappers and settlers from many miles around. The location may seem remote, but at a time when few roads existed and travel was mostly by canoe, the tavern was, in fact, easily accessible. Chevalier's numerous and comely daughters are said to have added to his tavern's popularity.[4]

In 1834, possibly in connection with the Ohio-Michigan boundary dispute, Ambrose Rice surveyed the area for the U.S. Government. His report describes the Cedar Point Marsh as "impassible," and indicates a lighthouse on the point.

According to an 1875 atlas, all the Lake Erie marshlands in Lucas County belonged to one John H. James, who may have been a government agent in charge of land sales.[5] These marshes were already becoming famous for some of the best waterfowl hunting in the country, and sportsmen had been buying up parcels for private clubs since the 1850s.[6] On March 25, 1882, the Cedar Point Shooting Club was incorporated by a small group of Toledoans: Miles Carrington, Peter Berdan, Robert and John Cummings, Joseph Secor, and Emery Potter. In later years Arthur Secor, Oliver Payne, and Widmont Ketcham were admitted. Club rules specified "no shooting in spring, on Sundays, or at night, and no betting on the premises."[7]

The club originally purchased 5,000 acres along the lakefront between Little Cedar Point and Ward's Canal, five miles to the east.

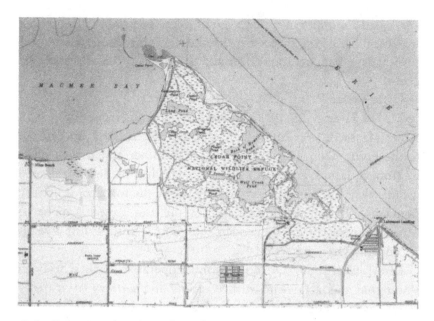

Cedar Point Marsh, U.S. Geological Survey, 1967. The remaining triangle of Lamb Beach appears just east of the Point, which has shifted south of its position on Terrell's 1917 map. The entire undiked section of shoreline marsh east of Wolf Creek Pond is now under water. The drained section shown in white at the west end of the refuge property was a favorite place for pheasant hunts.

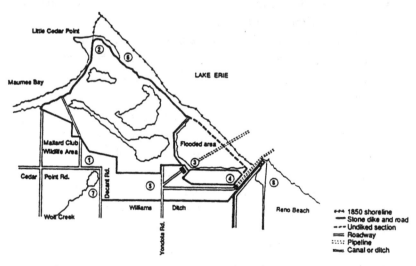

Cedar Point Marsh, 1990. (1) Entrance, Cedar Point Road. (2) Site of the Cedar Point Clubhouse. (3) City of Toledo water pumping station. (4) City of Oregon water pumping station. (5) Entrance to marsh, Yondota Road. (6) Site of Lamb Fisheries, until about 1940. (7) Goose Pond, the present end of Wolf Creek. (8) Anchor Point Marina.

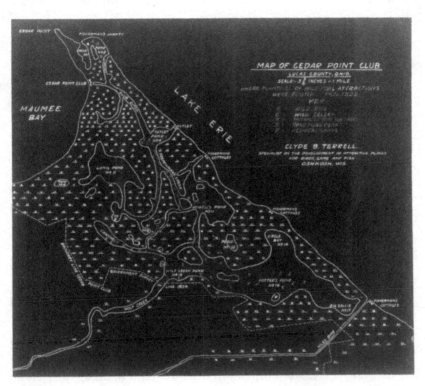

Cedar Point Marsh, drawn by Clyde Terrell, 1917. Terrell had been hired by the Club to plant wild rice and other favorite waterfowl food plants. Notice the outlet east of the Point. The entire beach, now mostly washed away, was at that time 50–250 feet wide. Courtesy of Tom Mominee.

In 1902, 900 acres on the eastern border were sold to Henry Reno and George Howard, setting the marsh's present boundary at Cooley Canal. A forty by sixty-foot, two-story, fifteen-room clubhouse was built on the bay near the Point. During its construction, members and workers lived in a barge drawn up on the beach. A porch and second-floor veranda covered all but the east side of the clubhouse, and the roof was surmounted by an observation platform, giving a fine view of the marsh and the bay.

The bay beach belonged to the Cedar Point Club, but the lake beach did not, having been deeded to Arthur Howell in 1876. Howell's original beach property extended from the Point all the way to Crane Creek, but over the years various parcels were sold to commercial fishermen for $1,000 a mile.[8] In 1889, the team of Clarence Lamb and Eugene Thompson bought the section of beach lying between the base of Little Cedar Point and Cooley Canal. Thompson and Lamb built a three-room house and several net sheds on a large triangle of this beach about three quarters of a mile east of the Point. A roadway and bridge across Cooley Canal made the fishery headquarters accessible by land.

Though today most of the beach is under water, in early days it was quite substantial. As late as 1915, E. L. Fullmer described both the bay and lake beaches as 50 to 250 feet wide, supporting not only many willow and cottonwood trees, but also sycamores, red maples, and dogwood. However, Fullmer also noted that even the highest parts of the beach were "but little above the reach of waves of violent storms."[9] The beach was already being eroded.

In 1928, the Cedar Point Club instituted a lawsuit against Lamb and Thompson, claiming that the firm's original beach property had been washed away. Clarence Lamb's son Edward argued the case for the fishery and won, but other lawsuits followed until the dispute was finally settled by mutual agreement in 1941. In 1943 the fishery buildings were destroyed by high water. Though most of the beach is now gone as well, at the site of the fishery the triangle of beach remains, covered by a tangle of fallen trees.

Over the years, Toledo members gradually gave up their membership to a group of M. A. Hanna Company executives from Cleveland. The last Toledoan to resign was Arthur Secor, who left the club about 1930. Secor was interested in birds, and in the late 1920s he began visiting my office to ask questions and share his findings. Through his influence I was given permission to visit the marsh to study bird life—a great concession at that time because of the fear of poachers, who illegally shot or trapped a great many ducks and mammals. All the marsh owners were plagued by poachers and tended to view any visitors, including birdwatchers, with suspicion. In exchange for free access to the marsh, I was to send reports of my findings to the U.S. Biological Survey (predecessor of the present Fish and Wildlife Service) and to the president of the club. Until the duties of raising a family slowed my activities in 1940, I visited the marsh about fifty times a year.

My first trip was on March 3, 1929. At that time, overland access was possible only by taking Cousino Road to Maumee Bay, then following a crude road east along the bay shore to the clubhouse. However, that same year a canal was dredged from Cedar Point Road north to Lake Erie, and within a few years the present two-and-a-half-mile stone road was built on top of the dike as far as the clubhouse. This canal provided drainage for a small area along the western property line, which was kept in crops or grass and used for pheasant hunts. Other canals were soon dredged throughout the marsh so that hunters and trappers—and I—could penetrate deep into its interior.

During the storm in the spring of 1929, Lake Erie broke through the earthen outer dikes and inundated marshes, subdivisions, and farms from Little Cedar Point east to Crane Creek. Even Cedar Point Road was flooded for a short time. In the drought years that followed, outer dikes were enlarged and rebuilt all along the lakeshore, not only to protect against future floods, but also to hold water in the drying marshes. When the Cedar Point Marsh continued to suffer from the relentless drought, the club installed a large electric pumping station near the mouth of Cooley Canal to pump in water from the lake.

Pumping was supervised by Louis Bono, a member of the Cherokee family that had given its name to the village of Bono. Louis, who was one of Frank Bunno's sons, lived with his family in a cot-

Lamb Beach

Separate ownership of the Cedar Point Marsh and its adjacent beach resulted in more than fifty years of conflict, bitterness and litigation. In 1980, Edward Lamb, who loved a good story, recorded some memories of those years from the fishermen's point of view:

> I spent two summers in 1910 and 1911 working in the fishing grounds of my father, Clarence Lamb. We had a two or three-room house or shack for the fishermen, and tarring centers and net repair shops. Each fisherman displayed his prizes on the wall—I remember eagles with wingspreads of twelve feet or better on the porches. We set out at four o'clock in the morning, placed our nets and then ran our 500 rod catline. We really brought in whoppers, like our 50-pound catfish and 40-pound pickerel. Around noontime we lifted the nets, pulled them ashore, placed the catch in boxes and hauled them off in the fishing boats, some seven or eight miles to our fish house on the Maumee River. In the wintertime we harvested ice from the Maumee River, stored it and packed the boxes for shipping off to urban markets.
>
> Many of the fishermen were Frenchmen, a rugged lot. . . . [They] were in sharp competition with the hunters and trappers and with each other for the use of the beach, and brawling and gunplay were not uncommon. Many nights when my father did not return home, he spent with a shotgun overlooking his fish nets. But these problems were as nothing compared to the efforts of the rich hunters from Cleveland, Ohio who had organized the Cedar Point Hunt Club and looked on these fishermen as dangerous trespassers. There were fist fights and court cases. . . .

tage on the beach a couple of miles east of the Thompson and Lamb fishery. The long drought brought at least one spring plague of tiny blackflies, which haunted the beach for weeks, drawing blood from any exposed skin. I was impressed by the resourcefulness of the sufferers, who tucked sheets of newspaper under their

My father sought only to retain fishing rights and ownership of the beach. However, we or "our boys" always carried on some hunting on the beach, especially of muskrats. Indeed, we took tens of thousands of muskrat off the beach. . . .

When I became a lawyer in 1928, I joined the firm of Edwin Marshall, [which] had been retained by the Cedar Point Gun Club to institute a lawsuit against their enemies, the pesky fishermen, and against our family as owners of the beach. I retained Lloyd T. Williams and Edward Ray to represent our family. We had an emotional, noisy trial that took many weeks; the rich hunters from out of town against the local working fishermen. Actually, the issue boiled down to a question of whether or not the beach had receded some distance, perhaps a mile or so, since my father bought it.

The jury even visited the premises. Only half of them were able to make it most of the way on the beach. It was wintertime. The beach was covered with snow and it turned out to be a rough journey. We crossed many areas where there was great concentration of trees and other vegetation. We had borings showing the age of these trees and some of them were well over 100 years old. . . . On that trip to the beach the jury was permitted by Judge Hahn to talk to all the participants—with all lawyers present. Possibly, we helped the jurors choose the side of the poor, local fishermen, who were being deprived of their livelihood by that very wealthy group of Cleveland corporate executives. . . .

The wind-up of the trial was quite a tearful event because the jurors wanted to boisterously demonstrate their sympathy with the fishermen. After all, we had paid the taxes and occupied the beach for many, many years. Now, the time had arrived to stop the harassment of the innocent local people!

Reprinted with permission from Edward Lamb, "The Grant of Lamb Beach to the Public," 1980.

trouser legs and down inside their socks, which are no barrier to these vicious pests.

About this time, the club began keeping a pen of clipped Canada geese near the clubhouse to attract migrating geese, the first of several projects designed to improve the already excellent hunting.

In later years, pheasants were raised in pens and released for fall hunting.

During Prohibition, the isolated point was a favorite spot for dropping off crates of beer and whiskey smuggled in from Canada. From there they would be loaded on smaller boats and taken inland. On one occasion, I found a seven-foot pile of beer crates on the beach, evidently waiting to be picked up. Another time, as I drove along a dike at the marsh border, I saw several men loading similar crates into a truck. I was a bit nervous, since in those days shootouts were not uncommon between smugglers and the Coast Guard up in Lake St. Clair, and I was obviously intruding on an illegal operation. To my surprise, the men waved at me casually and went on with their work. It may be that they recognized my old Ford, or even that they were marshmen doing a little moonlighting; in either case I thought it would be prudent to drive on by without looking too closely at their faces, and that is what I did.

In those years, travel in the marsh was easiest in a punt. During the 1930s and early 1940s I made many punting trips through the marsh, sometimes following a channel along the lake from the clubhouse all the way to Cooley Canal. Through the kindness of Edward Clomb, who owned a farm at the intersection of Cedar Point and Decant roads, I was able to follow Wolf Creek from that point to its mouth at Potter's Pond, deep in the marsh. Wolf Creek was cut off in the 1940s, when new dikes were built all along the inland border of the marsh, and now ends in Goose Pond, west of Cedar Point and Decant Roads.

When all these marsh canals were new and deep, they were full of fish, including northern pike, largemouth bass, black crappies, bowfins, and numerous bullheads. Over the next several decades, in spite of periodic dredging, the canals gradually became more shallow and the number of fish declined. Later still, when the Cedar Point Marsh became part of the Ottawa National Wildlife Refuge, the canals were no longer dredged at all; they filled up completely, and the fish disappeared except for carp and an occasional bullhead.

In 1941, the Toledo water pumping station was installed at the southeast corner of the Cedar Point Marsh, drawing water from Lake Erie for Toledo and points south. The agreement between the city and the club included a provision that the station would furnish all the water needed to maintain levels in the marsh. The city of Oregon pumping station was installed nearby in 1964.

Throughout its history, erosion of the outer barriers has been this marsh's greatest problem. Bordered on the east by Lake Erie and on the west by Maumee Bay, it is even more vulnerable to high water than are other Lake Erie marshes. In 1875 the beach about a mile and a half east of Little Cedar Point was already broken by a deep channel called Norman's Strait. This opening was still there in 1915, reportedly some 300 feet wide where it entered the marsh.[10] It appears on the 1917 map (p. 121), but by 1922 drifting sands had filled it in. Between 1929 and 1965 I saw the outer dikes destroyed three times by high lake levels, and each time the new barriers had to be built further inland. The Little Cedar Point sandbar shifted at least a quarter of a mile south of its 1929 position, and in 1973 it disappeared under the high water. By 1990 it was almost entirely washed away. During the last three decades, I have also seen many shoreline willows over a hundred years old uprooted by the high water.

Until the late 1950s, the five-acre tip of the Cedar Point Marsh peninsula was one of the greatest bird concentration points in the entire Toledo area. Spring and fall migrants followed either the tree-lined shore of Lake Erie or the wooded southern border of the marsh and collected at the Point, reluctant to fly over the open lake or Maumee Bay. Dragonflies, bumblebees, and monarch butterflies also appeared on the Point, sometimes in great numbers.

On various trips to the Cedar Point Marsh during the 1930s and 1940s, I saw a million bank swallows, 350 bluejays, 250 hermit thrushes, 300 Swainson's thrushes, 300 golden-crowned kinglets, 200 Tennessee warblers, 700 yellow-rumped warblers, 200 blackpolls, 50 Wilson's warblers, 110 redstarts, 300 indigo buntings, 350 white-throated sparrows, 300 swamp sparrows, and 500 song sparrows. Unfortunately, when trees were destroyed by erosion along Lake Erie and cut down along the marsh's inner border, birds were no longer funneled toward Little Cedar Point in unusual numbers. Today, the only known concentrations occur in the Navarre Marsh and along the bird trail in the Magee Marsh.

From 1929 through 1939, common terns nested on the sandbar of the Point (see Chapter 3). They reached a high of 5,000 before abandoning the site, driven off by erosion and the presence of seine fishermen. A few pairs of piping plover bred there from 1932 until 1943; then they, too, deserted the place. The marsh itself has, of course, always been a haven for water birds of all types. Some of the oldest marsh guides told me that at one time a small colony of black-crowned night herons nested there.

On one of President Eisenhower's visits to the Cedar Point Marsh, the club scheduled a pheasant hunt on the grassy area at its western boundary. Unfortunately, flushing the birds from the tall grass turned out to be extremely difficult, raising the possibility that the President might go home pheasantless. With the club's reputation at stake, Superintendent Cornelius Mominee ordered a trail secretly mowed through the grass. On the day of the hunt, a number of guides were stationed beside the trail at intervals, well-hidden in the heavy cover. As the President passed by, Mominee would call out, "There's one . . . *now!*" and a guide would toss a squawking bird into the air. Needless to say, the trip was a great success, and if Ike suspected "fowl play," he never let on.

Early in 1963, plans were inaugurated to sell the Cedar Point Marsh for industrial development—there was talk of a steel mill—and Lake Erie Harbors was incorporated for the purpose. Fortunately for all Ohioans, this sale never took place. William Sherwin, a conservation-minded club member, inspired the decision to donate the marsh to the North American Wildlife Foundation. On December 18, 1964, Joseph H. Thompson, President of the Cedar Point Club, presented to the foundation the marsh and clubhouse, appraised at one million dollars. The only provision was that it could not be turned into a public park, campground, or picnic area. C. R. Gutermuth, Secretary of the Foundation, turned the property over to Secretary of the Interior Stewart Udall on January 22, 1965. The Cedar Point Marsh thus became the Cedar Point National Wildlife Refuge, a part of the Ottawa National Wildlife Refuge comprising 90 acres of cropland, 10 acres of upland, 1,700 acres of marsh, and 445 acres of water.

Shortly after the government assumed control, plans were drawn up to renew the outer dikes once again, a project that was accomplished in the late 1960s. Clay for the base was taken from a large borrow pit beside the marsh road. The barrier, reinforced with massive chunks of limestone, measures sixty feet at the base, fifteen feet across the top, and fifteen feet in height. Because of the great expense, estimated at $2.5 million, the shoreline northeast of the Toledo Pumping Station was not diked. Instead, a smaller dike was built closer to the southern boundary of the marsh. This decision resulted in the loss of a square mile of marshland when the smaller dike failed to withstand the high lake levels of 1972 and 1973 (see map on p. 120). The lost marshland may yet be reclaimed, if reasonably low water levels ever return.

Originally, planners had intended to use the old clubhouse as a National Waterfowl Museum, but they had not reckoned with vandals. The building was readily accessible to boaters, and in a comparatively short time its interior was virtually destroyed; worse, arsonists had already burnt several refuge buildings. In order to prevent an uncontrolled fire at some future time, in April of 1972 the historic old clubhouse and the caretaker's home were burnt to the ground.

Since 1981, area birder Joe Komorowski has led a bird count in the marsh on the middle Sunday of each month. The primary aim of the Refuge is to preserve the marsh habitats and their wildlife undisturbed; for this reason, all other birding trips must be authorized at the Ottawa Refuge headquarters. In addition to the Sunday counts, each fall biologists from the Crane Creek Wildlife Experiment Station at the Magee Marsh conduct aerial counts of waterfowl in all the Lake Erie marshes, including the Cedar Point Marsh.

Ottawa National Wildlife Refuge

The Ottawa National Wildlife Refuge was established in 1961, when the Fish and Wildlife Service bought a number of privately owned small marshes between the state-owned Metzger and Magee Marshes. The original 4,807 acres included 2,500 acres of marsh,

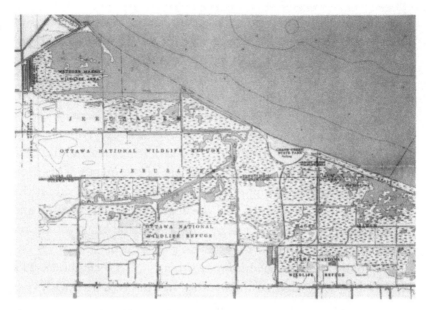

Ottawa National Wildlife Refuge, U.S. Geological Survey, 1967.

2,000 acres of cropland, and 125 acres of timber, with another 187 acres of canals, ditches, and roadways. Over the next six years, more land was added to the refuge until today the refuge complex comprises 8,315 acres in four townships.

The birth of the Ottawa Refuge was a difficult one, requiring the concerted efforts of state and federal authorities and scores of Ohio conservation organizations. In 1958, planners with the Ohio Division of Wildlife began discussing the advisability of a federal wildlife refuge in the Lake Erie marshes. Over half the fall peak numbers of waterfowl in the state were found in these marshes—typically, about 61,000 birds in mid-November.[11] It was vital to preserve the best remaining marshes, since so many had already been irrevocably lost, including Toledo's Detwiler Marsh, now a golf course, and the Raisin River wetlands in Michigan, piled high with fly ash.[12] And although Ohio sportsmen purchased many duck stamps to finance federal refuges in other states, including over 100,000 acres in Michigan, Ohio still had only the 77-acre refuge on West Sister Island. The proposed refuge, first called the Erie National

The Marshes of Southwestern Lake Erie

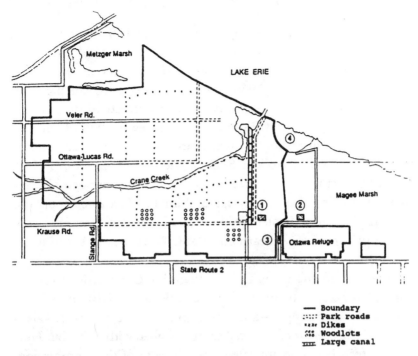

Ottawa National Wildlife Refuge. Notice that a narrow strip of the refuge lies between State Route 2 and Magee Marsh. (1) Refuge headquarters, parking lot, entrance to hiking trails. (2) Magee Marsh Wildlife Area headquarters. (3) Crane Creek State Park headquarters. (4) Crane Creek Park.

Private marshes purchased to create the Ottawa National Wildlife Refuge, July 28, 1961. (1) Continental Marsh. (2) Roy J. Dewey Marsh. (3) Douglas Marsh. (4) Eisenhour Marsh. (5) France Marsh. (6) Pintail Marsh. (7) Hunter Marsh. (8) Ritter Marsh. (9) Searles Marsh. (10) Willow Point Marsh.

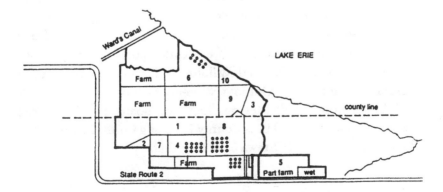

Wildlife Refuge, was approved in 1959 by the U.S. Regional Office Land Acquisition Review Committee and the Central Office Review Committee of the U.S. Fish and Wildlife Service.

Suddenly, on June 29, 1960, a bombshell exploded: the president of Horizon enterprises in Wooster, Ohio announced that in 1959 his firm had purchased Pintail Marsh, a key portion of the proposed refuge, intending to turn it into a real estate and marina development called Holiday Shores. Although fully a third of the 473-acre tract was under water at the time, plans included 778 residential building sites, marinas, a bathing beach, a 2,000-foot landing strip for aircraft, steel jetties on Lake Erie, community wells, and a private sewage treatment plant. Work was to begin in the spring of 1961.

The Toledo and Lucas County Planning Commissions rejected this plan because of inadequate sewage disposal facilities and a minimal water supply—a decision proved wise by the floods of 1972–73. Conservation-minded citizens sighed with relief, but Horizon Enterprises soon appealed the judgment of the Commissions. At that point, federal, state, and local organizations joined forces and swung into action.

The U.S. Fish and Wildlife Service began an immediate campaign to acquire Pintail Marsh and other key lands and to form a federal wildlife refuge before any other unwelcome developments could arise. For its part, even before the Planning Commission hearing, the Ohio Wildlife Council had been urging the state to purchase Pintail Marsh at once and hold it until federal funds became available.

On the local level, in 1961 seven area associations were consolidated into what ultimately became known as the Northwestern Ohio Natural Resources Council.[13] The Council began a letter-writing campaign, joined by the Ohio Division of Wildlife, Izaak Walton League of America, League of Ohio Sportsmen, Ohio Conservation Congress, National Wildlife Federation, Ohio Natural Resources Commission, Outdoor Writers of Ohio, Toledo Naturalists' Association, Toledo Metropolitan Park Board, Toledo District Garden Forum, Toledo–Lucas County Planning Commissions, Ohio Biological Survey, and Ottawa Shooting Club.

The campaign brought results. The appeal by Horizon Enterprises was denied, and the Ottawa National Wildlife Refuge was formally approved by the Migratory Bird Conservation Commission on March 15, 1961; it was officially launched on July 28 of that year.

In addition to Pintail Marsh, the refuge took over nine other established private shooting marshes. These clubs were, in a sense, losers in the conservation campaign, since their members no longer had a place of their own for hunting. Furthermore, it was through these owners' efforts that the wetlands had been preserved or restored. Duck clubs and their members should forever be regarded as the unsung heroes in the battle to preserve the Lake Erie marshes.[14]

The first manager of the Ottawa Refuge, Alfred Manke, took on the task of establishing a management strategy for this mosaic of land parcels, each with its own history. Some marshes had been preserved virtually unchanged. Some, like Pintail, had been drained for farming and then reflooded. The 197-acre Continental Marsh was originally not a marsh at all, but swamp forest drained on its south boundary by Crane Creek; the land was cultivated until 1947, then purchased by the Continental Marsh Shooting Club, which pumped in water from Crane Creek to create the present cattail marsh. About two square miles of farmland lying between Veler and Ottawa-Lucas roads was kept in crops or pasture and only reflooded when Lake Erie accomplished that task during the storm of 1973. Other farmlands included in the original purchase still remain under cultivation, providing grain to feed the masses of waterfowl that winter in the refuge.

The legacy of multiple ownership is displayed in refuge maps: in contrast to the unmarked expanse of the Cedar Point Refuge, the map of Ottawa shows a complex pattern of roads, canals, and ditches inherited from its former owners. Building on this patchwork legacy, Manke supervised construction of many more miles of dikes that now separate the marsh into small, manageable units. In 1965, this project employed between one hundred and two hundred young people from the Ottawa (County) Job Corps Conservation Unit at Camp Perry Center.[15]

Planning a Wildlife Refuge

The prime objective of all American National Wildlife Refuges is to preserve and create nesting and feeding habitat for migratory birds and endangered wildlife. Yet the needs of other species, including human beings, must be considered in the balance. To do this, the U.S. Fish and Wildlife Service develops a Master Plan for long-range guidance of refuge management.

It begins with a detailed look at the refuge resources: an inventory of the lands and waters, climate, soil and vegetation, native and migratory animals, and the existing and potential human impacts on the environment—roads, industrial areas, and nearby population centers.

The next step is to analyze the refuge for potential uses. Twenty uses, or "outputs," are being considered for the Ottawa National Wildlife Refuge Complex. Many of these uses compete with each other for space and time, so they must be brought into harmony with each other and with the goals mandated by the federal laws that established the National Wildlife Refuge System. Of course, not every acre of a refuge is ideal for all uses; wetlands make poor hiking trails, and ducks don't nest on sandbars.

The Fish and Wildlife Service uses a system for resolving conflicts between uses. It assigns a numerical value, based on national priorities, for each use. This is called the Refuge

Within a few years of the original purchase, the refuge acquired additional marshlands outside the original boundaries—including, in 1965, the Cedar Point Marsh. In 1966, the 591-acre Navarre Marsh complex was purchased for the refuge; however, the next year this plan was unexpectedly changed when the Toledo Edison Company and the Cleveland Electric Illuminating Company bought 520 acres of Darby Marsh near Port Clinton as a site for a nuclear power plant.

Benefit Unit, or RBU. Wildlife production usually has a high RBU, while environmental education and recreation are lower on the scale. For example, each goose produced may be valued at 800 RBUs; each hour of trail use at 75 RBUs. A conflict may arise when an area is suitable for several uses that are incompatible with each other. Wildlife refuge planning teams work to find a balance that provides an optimum mix of appropriate uses for the refuge.

Throughout this process, many people are invited to express their concerns: hunters, fishermen, farmers, educators, environmentalists, local agencies, land owners, and other groups. These concerns are weighed along with the needs of wildlife, the land resources, the environmental impact, and the objectives of the refuge system. Brought into balance, these elements become the Master Plan for each refuge.

The Master Plan explains how the refuge lands will be developed and managed; where facilities will be located; where unique historical, natural, or archaeological features will be protected; where hunting, fishing, and other recreation will be enjoyed; where lands and waters will be managed to improve wildlife habitat. Most important of all, the Master Plan sets the objectives for the refuge, defines the balance between man and nature on the land, and sets a standard by which to measure progress.

—*From the Technical Report on the Ottawa National Wildlife Refuge Complex, 1979.*

Because the underlying limestone at Navarre Marsh was closer to the surface and the beach ridge was higher there, Navarre was a much more suitable location for the power plant than Darby. The power companies offered not only to exchange Darby for Navarre, but also to rebuild the outer dike at Darby and to turn over the management of the 350-acre marsh in the Navarre complex to the Ottawa Refuge, together with an additional 100 acres purchased from the owners of Sand Beach Marsh. This agreement, certified

October 4, 1967, gave the Fish and Wildlife Service control of 1100 acres, nearly 600 more than the original purchase. The refuge continues to acquire land, with a goal of 5,000 more acres by the year 2000.

From the start, trapping has been permitted in both state and federal marshes from the end of waterfowl season to March 15 of each year. Studies have shown that in a managed marsh, with its minimum of natural enemies and ideal water levels, the muskrat population increases to the point where the animals are susceptible to a form of rabies. To prevent this catastrophe, trappers are allowed to submit sealed bids for assignment to one of seventeen designated trapping units. During the last five years, fifty trappers have taken an annual average of 6,075 muskrats. Raccoons are in second place with a catch of seventy-five per year.

Waterfowl hunting is restricted to four days a week in a few selected areas of the refuge. No other hunting is allowed. Originally only geese could be taken, but ducks were added in 1986; limits are set each year by federal regulations, and steel shot is required. About 500 hunters, chosen by lottery, bagged an average total of 150 geese each season from 1986 through 1990.

Public access to the refuge is limited to nearly ten miles of hiking trails winding through a variety of habitats. In 1971, Toledo naturalist Robert Crofts began leading a monthly bird count on these trails, an important source of information that Edward Pierce and Chris Crofts have carried on since Robert's death in 1980. Their observations, and those from other counts in the Darby and Cedar Point Marshes, are the only sources of information on bird populations in these marshes.

Navarre Marsh, also called the Davis-Besse Marsh, is bounded by the Toussaint River, Lake Erie, and Sand Beach Marsh. It was owned by the Navarre Shooting Club for nearly 130 years, from 1838 until its purchase by the power companies in 1967. Between the west boundary of the property and the lake is a solid limestone ridge, now the site of the Davis-Besse Nuclear Power Plant. Protected from Lake Erie floods by this natural barrier and carefully

By the Numbers

The monthly bird counts taken in the Ottawa National Wildlife Refuge are a valuable source of both historical and current information on bird populations in the marshes. Records are kept on cards, one for each species, showing at a glance the frequency of its appearance at each season as well as year to year. Below are two samples, kindly supplied by Edwin Pierce.

Great Blue Heron – ONWR

	80	81	82	83	84	85	86	87	88	89	90	91	92
Jan	✕	19	9	22	37	45	3	39	42	31	3	10	19
Feb	✕			5		10	5	16	5	5	1	47	40
Mar.		1		5	7	1	1	10	40	7	16	29	31
Apr	14	100	7	45	14	45	150	120	120	125	87	86	135
May	35	53	26	80	24	100	50	78	55	125	82	65	115
June	25	75	65	52	6	100	265	100	138	20	95	101	151
July	38	78	25	50	63	60	70	85	430	116	100	157	82
Aug.	60	40	135	25	40	46	50	54	147	93	129	134	46
Sept	22	94	43	21	40	24	92	49	175	92	94	44	
Oct.	35	35	28	25	45	26	69	54	100	93	76	88	
Nov.	19	17	19	7	21	46	50	31	20	85	75	36	
Dec	6	11	14	18	10	37	20	30	112	61	48	86	

Common Yellowthroat – ONWR

	80	81	82	83	84	85	86	87	88	89	90	91	92
Jan.	✕					1							
Feb.	✕												
Mar.													
Apr.													
May	2	3		1	4	6	1	2		18	20	10	7
June	5		37	13	20	12	7	5	21	12	10	13	48
July	3	4		3	32	16	12	25	14	17	19	27	31
Aug.	4	15	10	5	36	9	15	5	10	18	9	6	16
Sept.	3	3		1			5	10	6	5	5	4	
Oct.	3	1	4	5	3	3	1	3	3	6	2	4	
Nov													
Dec	1				1				2			1	

managed, Navarre is one of the least disturbed marshes along Lake Erie.

Not only is it undisturbed, but packed into Navarre's 535 acres are all the habitats associated with a marsh complex. Between the cattail marsh and Lake Erie lie a long, narrow pond, an extensive growth of buttonbush, and a border of trees, some of them unusually large hackberries. Marsh grass grows along the edges of the cattails. The shelter of the wooded area and an abundance of bird food, especially white dogwood berries and pokeberry, combine to make this marsh northwestern Ohio's major gathering place for migrating birds. The presence of the 493-foot cooling tower and other structures does not appear to disturb the wildlife in the least.

Twice a year, state wildlife biologists set out banding nets in the wooded border to capture a representative selection of migrating birds (see pp. 64–67). The extent of these migrations is clear from the spring figures: in 1989, banders Mark and Julie Shieldcastle netted 6,859 individual birds representing 101 species. In 1990, the

Navarre Marsh, U.S. Geological Survey, 1971. Sand Beach Marsh lies to the west, Toussaint Marsh to the east across the Toussaint River.

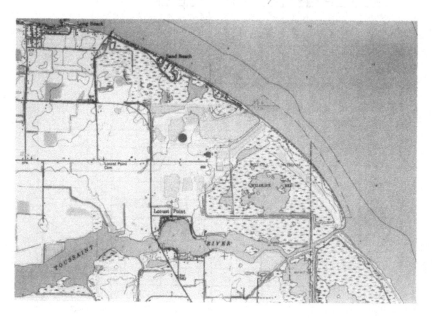

total was 7,914 individuals representing 109 species. The most abundant species in both years were the magnolia warbler (699 in 1990), white-throated sparrow (669 in 1989; 538 in 1990), and yellow-rumped warbler (612 in 1990). The rarest bird netted (on May 11, 1990) was a clay-colored sparrow.

Darby Marsh comprises 604 acres in Erie Township, Ottawa County, west of Port Clinton between State Route 2 and Lake Erie. According to a 1900 atlas of Ottawa County, 118 acres of the property were owned by the Lake Shore and Michigan Southern Railroad and 200 or more acres by the Lacarpe Shooting Club.[16]

In 1958, Darby was owned by Ethyl Hazelton and J. W. Galbreath, with approximately 354 acres of shooting grounds. In 1968, the Davis-Besse interests purchased 520 acres of the marsh and turned them over to the Ottawa National Wildlife Refuge in exchange for Navarre Marsh. Because of high Lake Erie levels, all buildings were removed and salvaged in 1973. Bird populations in this marsh are monitored by John Redman.

Throughout the years, conditions in Darby marsh have varied greatly. At one time, for example, Lacarpe Creek had two outlets into Lake Erie. At another time, much of the marsh was a large open bay extending inland. Since 1975, stone outer dikes have stabilized the marsh.

The struggle over **Pintail Marsh** in 1960 helped focus public attention on the threat to marshlands all along the western Lake Erie shore. As we saw in the previous chapter, Pintail was one of the first of the low-lying marshes to be drained for agriculture in the early 1900s. After a flood in 1910, its owners rebuilt the earthen dikes and reclaimed the land, but they were decisively beaten by the storm of 1929. With half its acreage under water, Pintail Farms was abandoned and it reverted to marshland.

Local sportsmen established a duck hunting club and enjoyed excellent shooting for the next two decades, until they began to neglect the pumps. The resulting low water levels encouraged the growth of cottonwoods and willows in the marsh, and it deteriorated rapidly. It was probably this low water that led Horizon Enterprises to believe that the marsh could be drained entirely and turned into a recreational subdivision. After its incorporation into the Ottawa Refuge, Pintail was flooded by Lake Erie once more, in 1973. Today it is protected by high stone dikes installed in 1983–84.

Magee Marsh Wildlife Area

The vast, open expanse of the 1800-acre Magee Marsh lies between Turtle Creek and the eastern boundary of the Ottawa Refuge. Of all the marshes in this book, Magee is the most accessible to the general public; in fact, one of its primary goals is the accommodation of visitors with widely varying interests. The marsh has always been well known for its excellent hunting and trapping and for the variety of its waterfowl. Magee Marsh is divided into two sections: The seventy-two-acre Crane Creek State Park, with bathing beach and picnic area, is managed by the Ohio Division of Parks and Recreation, while the Division of Wildlife manages the 2,131-acre Magee Marsh Wildlife Area.

The wildlife area supports a variety of recreational and scientific uses. For anglers, there is a fishing and boat launching site on Turtle Creek. Canada Goose Day is a popular annual spring event. The two-story Sportsmen Migratory Bird Center is a visitor center and management headquarters, featuring a display of antique waterfowl decoys and an educational collection of mounted specimens of birds that visit the area. Nearby is an observation tower for studying water birds. Magee's most famous attraction, however, is the elevated bird trail laid out in a small woodlot not far from the beach; for some reason, huge numbers of migrating birds pass through this small area in both spring and fall, attracting thousands of people with binoculars and cameras. One reason for the popularity of Magee is that no fees are charged for either bathing or birding.

In 1900, the land adjoining Crane and Turtle Creeks was owned by the Crane Creek Shooting Club, Rocky Ridge Shooting Club,

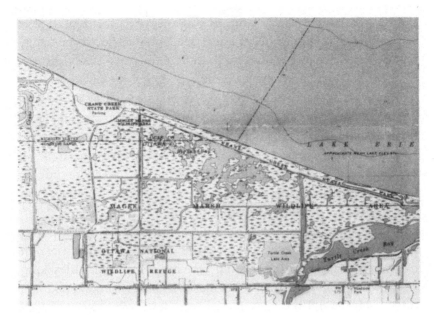

Magee Marsh, U.S. Geological Survey, 1967.

Magee Marsh. The large woodlot at the south boundary, near Turtle Creek, was the original site of the bird trail. (1) Crane Creek State Park, 72 acres. (2) Bird trail. (3) Headquarters, Magee Marsh Wildlife Area, 2131 acres. (4) Headquarters, Crane Creek State Park. (5) Fishing area and boat launching site on Turtle Creek. (6) Headquarters, Ottawa National Wildlife Refuge.

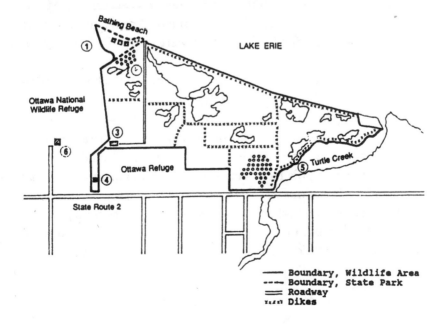

and Frank Stang.[17] Club members rode in trains to the village of Rocky Ridge, then travelled the rest of the way by horse and buggy. In 1903, a Cleveland group purchased 1,300 acres west of Turtle Creek[18] and hired John Nicholas Magee to dredge out canals and shooting holes for them. Magee's dredge, operated by a five-man crew, was forty feet long and twenty-four feet wide. He did a good deal of drainage work for neighboring owners, and when, as sometimes happened, they were unable to pay the costs, he took his payment in marshland. In this way he acquired 1,400 acres adjoining the Cleveland Club property, and in 1904 he purchased the club's holdings as well, increasing his land to 2,700 acres.

At the time, many of the wetlands west of Turtle Creek were being drained for farming, and Magee decided to follow suit. He excavated a drainage canal, nine feet deep and four miles long, along what is now the road to the bathing beach. But, like so many others, Magee did not reckon with the destructive force of Lake Erie; his land soon flooded, and he prudently gave up farming in favor of hunting and trapping on a large scale.

This decision marked a bold departure from previous attitudes toward marsh management. In the past, owners had generally followed one of two courses: they either preserved their marshes for their own pleasure, or drained them and sold or farmed the land for profit. But John Magee preserved his marsh *and* turned a profit, by transforming the marsh itself into a business enterprise. Muskrats and raccoons were sold for fur, while ducks and geese, turtles and frogs were sold for food. Hunters were charged a fee to shoot in the marsh.

This last practice signaled yet another important change. Members of the big shooting clubs, who paid large sums to buy and maintain their marshes, tended to be wealthy men—before the depression, for example, Cedar Point Club members were all millionaires. Now, for the first time in decades, a hunter of modest means could, after paying his fee, freely set up his blind or moor his boat at dawn in a frosty marsh.

When cattails and other vegetation invaded his property, Magee increased the depth of water in the marsh to change the amount and type of vegetation, hoping to find the most attractive habitats for ducks and muskrats. He thus became one of the first marsh owners to practice systematic, purposeful management techniques. These techniques included at least one that is now illegal: before live decoys were banned, Magee maintained a flock of three hundred call ducks, and his marsh soon became very popular with hunters.

After Magee's death, his daughters Julia, Ruth, Catherine and Sadie carried on the work. All forms of wildlife were harvested in season. Two trappers once captured 2,800 pounds of snapping turtles, which were sold for two cents a pound. On another occasion, fifty-seven dozen bullfrogs were taken in one night. At the height of the fur season in 1919–20, muskrat hides alone brought $40,000. These profits attracted many poachers that winter, and twenty-five of them were arrested in a single night. Poachers were always a problem, catching frogs, hunting at the edges of the marsh, or sneaking traps into the marsh itself.

I first visited Magee Marsh on June 2, 1928, with author and birder Nevin O. Winter, who had done legal work for the Magee sisters. We rowed a small boat down the main canal from the clubhouse to the Lake Erie beach and found a heronry housing twenty pairs of black-crowned night herons, along with eight pairs of great blue herons. We also saw a prothonotary warbler, then considered rare. The following year, on June 22, we found that great blue herons had begun to dominate the heronry: there were now fifteen pairs of night herons and twenty-five pairs of great blues. Shortly after that I was granted free access to the Cedar Point Marsh, so Winter and I did not visit Magee Marsh again. It is interesting to note that, although Winter was the family lawyer and I a harmless birder, a man in a boat was assigned to follow us wherever we went—such was the Magee sisters' fear of poachers. Today their beloved marsh welcomes many thousands of visitors each year, more than all the other Lake Erie marshes together.

Magee Marsh was rented by a private club for many years, but the cost per bagged bird increased until it reached about fifty dollars, and the renters abandoned the project. In August, 1951, the marsh was purchased by the Ohio Department of Natural Resources, which in 1956 established the Crane Creek Wildlife Experiment Station as a center for research on wetlands and their wildlife.

Funds for these studies come from the sale of hunting and fishing licenses.

A major goal of the experiment station is to develop habitats that attract and nourish marsh wildlife, particularly waterfowl—a goal very much in the spirit of John Magee. In practice, this means learning when and how much to change the water level of the marsh in order to foster the growth of waterfowl food plants. In a sixteen-year marsh management study begun in 1957, twenty-two miles of dikes divided the marsh into seven major units, each with its own water control. About five hundred acres of wet meadows and timber were left untouched, while water in the other six units was adjusted to various levels at various times of the year.

In general, the study determined, standing water will kill cattails, cane, and most woody plants in one year. A second year at a minimum water level encourages the growth of duck food plants. The unit is then reflooded before the fall migration arrives looking for open water. Each unit is still managed separately, and water levels are varied from one to another so as to attract the widest possible range of wildlife. Waterfowl augment their natural food with grain from surrounding farms. Twenty-nine species of ducks, five species of geese, and two species of swans have been identified in and near Magee.

As a result of this research and the careful practice of management techniques arising from it, the harvest rates for public waterfowl hunting at Magee are two to three times the national average. The Wildlife Area was opened to public hunting on a statewide lottery system in 1959, with twenty-four available blinds, and boats if needed. In 1976, thirty-two more blinds were installed in the adjoining Ottawa Refuge. The accompanying chart compares results of private and public hunting in Magee; it shows the vast difference in skill between the average duck club member and the occasional hunter. It also clearly shows that current duck harvests are not excessive.

Although hunters in the Ottawa Refuge may shoot both ducks and geese, those in Magee are allowed to shoot only ducks. The reason is Magee's Canada goose rearing project, begun in 1967 and now one of the most successful goose projects in the nation. (This

Waterfowl Hunting in Magee Marsh
under Private vs. Public Control
Average Annual Harvests

	Private Ownership	Public Ownership		
	1941–48	1951–56	1966–72	1984–88
Wood Duck	10	34	14	35.4
Green-winged Teal	42	64	100	81.2
Am. Black Duck	972	290	120	44.4
Mallard	1220	588	286	275.4
Black x Mallard	—	4	3	7.3
N. Pintail	606	66	170	40.8
Blue-winged Teal	36	9	17	15.8
N. Shoveller	29	27	65	41.4
Gadwall	62	116	36	43.4
Am. Wigeon	156	161	52	20
Canvasback	(*)	3	1	—
Redhead	(*)	5	15	2.2
Ring-necked Duck	(*)	1	36	8.2
Lesser Scaup	(*)	7	9	4.1
White-winged Scoter	(*)	—	—	1 only
Common Goldeneye	(*)	3	1	1.1
Bufflehead	(*)	3	4	6
Hooded Merganser	(*)	17	8	20.2
Common Merganser	(*)	9	2	1 only
Red-breasted Merg.	(*)	1	1	1.25
Ruddy Duck	(*)	2	14	3.8
American Coot	—	75	86	30.5
Canada Goose	4	13	protected	
Snow Goose	5 (1948)	0.25	protected	
Total Waterfowl	3149	1498	1041	673.3
Total Hunters	17 + guests	1210	540	1056
Season Length	41 days	58	44	22
Birds Per Hunter	184.8	1.2	1.9	0.6

Statewide, the breakdown on ducks bagged is: mallard, 30%; wood duck, 29%
divers, 12%; black duck, 11%; green-winged teal, 9%; all others, 5% or less.
At Magee, 41% of the ducks taken are mallards, followed by green-winged
teal at 12%.

(*) Through the years 1941-48, private marsh hunters bagged only seven of
these diving ducks, probably because they preferred puddle ducks.

project is discussed in Chapter 3, pp. 173–174.) It illustrates the second major goal of the Wildlife Experiment Station: the study of marsh wildlife. In addition to keeping track of the waterfowl harvest each year, Crane Creek biologists conduct aerial counts of waterfowl in all the surrounding marshlands and surveys of fur-bearing animals including muskrat, raccoon, beaver, and mink. One of their most important activities is the banding of songbirds, waterfowl, and eagles.

A determined effort has raised the number of eagles in the Lake Erie marshes to its highest in my lifetime: in 1990, fourteen nests successfully hatched ten eaglets (see Chapter 3, pp. 158–162). However, even though a young eagle has little to fear from predators, its first year is full of new experiences for which miscalculation could mean injury or death. Magee biologists keep track of young eagles by means of a solar and battery-powered radio transmitter attached to a light, simple harness on the eagle's back. Just before it is ready to leave the nest, each fledgling is fitted with a transmitter, then leg-banded and wing-tagged. Thereafter, a listener can not only locate the bird, but tell from the pattern of transmitted signals whether it is resting or in flight.

Transmitting range is two miles ground-to-ground, twenty miles ground-to-air, and up to fifty miles air-to-air. The system has so far revealed that the young eagles in northwest Ohio congregate only near the mouths of Crane Creek and the Sandusky River, apparently to avoid human contact. In the next few years, with the cooperation of biologists in neighboring states, northwestern Ohio's eagles may finally reveal their migration route.

The quarter-mile Magee Marsh bird trail is located in a fringe of trees alongside the wetlands in the northwest corner of the wildlife area. When the marsh was first acquired, a bird trail was laid out at the southeast corner of the property, adjoining Turtle Creek. The area seemed ideal—wooded, with bordering marshland and stream—but the birds ignored it.

Then, in 1963, game protector and naturalist Laurel Van Camp discovered that migrant birds were congregating in the northwest corner of Magee, in a seven-acre plot with only about 150 trees.

Red-tailed Hawk in Swamp Forest. *Credit: Tom Anderson.*

Nest of a Red-headed Duck. Magee Marsh, May 30, 1961. *Credit: Ohio Division of Wildlife.*

Red-spotted Purple Butterfly on grape leaves. *Credit: Lou Campbell.*

Painted Lady Butterfly. *Credit: Charles D. Wilson.*

Swan in cattails. *Credit: John Arnold.*

Great Egret. *Credit: Charles D. Wilson.*

Green Heron. *Credit: Lou Campbell.*

Great Blue Heron. *Credit: Charles D. Wilson.*

Muskrat eating cattails. *Credit: Lou Campbell.*

Canada Geese. *Credit: Charles D. Wilson.*

Female Common
Whitetail Dragonfly.
*Credit: Charles D.
Wilson.*

Spring Wooded Marsh. *Credit: Lou Campbell.*

Aerial photo of Magee Marsh and the Sportsmen Migratory Bird Center. *Credit: Ohio Divison of Wildlife.*

A corner of Navarre Marsh. *Credit: Mark Witt.*

Purple Loosestrife. *Credit: Charles D. Wilson.*

Great Bur Reed. *Credit: Charles D. Wilson.*

Why the birds gather in this particular area is a mystery, but a new trail was promptly laid out there and has remained popular with both birds and birders ever since. A variety of habitats attract so many different kinds of birds that on good days it is not unusual to see over seventy-five species. The trail is visited by over 35,000 people a year.

Birders are notoriously impatient with rules, and in the past, ignoring signs and fences, they wandered all over the area in pursuit of birds, trampling down the vegetation. The entire habitat began to suffer. To make matters worse, the trail is near the undiked bathing beach at Crane Creek Park, where a stiff northeast wind can easily sweep lake waters over the beach and onto the trail. After an April storm in the high water year of 1980, I found the first section of the bird trail under two feet of water. Two weeks later it was still flooded, and a family of hooded mergansers was swimming about the entrance. The trail remained muddy well into May.

The challenge was clear: how to make the bird trail restricted but accessible, without diking the beach. Biologists Mark Shieldcastle and Denis Case decided on a mile-long, elevated boardwalk. Financed by a state income tax check-off, the boardwalk was dedicated April 30, 1989 and soon fulfilled its original purpose: by midsummer of the first year the surrounding area was covered to a good height with jewelweed and other vegetation—and birders' feet remained dry. On my first trip over the boardwalk I discovered another of its benefits, as I stopped to talk to a young man in a wheelchair with bird book and binoculars, as able a birdwatcher as anyone there.

Metzger Marsh Wildlife Area

Because of its accessibility, the 558-acre, state-owned Metzger Marsh Wildlife Area is in use the year round. It is bounded roughly by the Ottawa National Wildlife Refuge on the south and east, by Lake Erie on the north, and by Ward's Canal on the northwest. A paved road on its west and northwest boundary ends in a public fishing pier that extends out into Lake Erie at the mouth of Ward's Canal. A popular launching site for small boats lies near the marsh's southwest corner. In winter, anglers use the end of the road and the parking lot to gain access to the lake for ice fishing. Bird-watchers scan the marsh from the road, especially in early spring. Because the area is one of the few places open to duck hunters on a first-come, first-served basis, it is extremely popular during the shooting season.

Like Pintail Marsh, Metzger Marsh was diked and drained in the early 1900s, becoming Metzger Farms, with two large houses and some barns beside a stone road near the center of the property. Along with Pintail it was first flooded in 1910 and then devastated in 1929, when the outer dikes were breached by a severe storm and the land was completely inundated. Rather than repair the dike, the owners abandoned the land, and the buildings gradually collapsed. During the low water of the 1930s, when marsh conditions were excellent, a duck hunting club operated the property. In 1955 it was bought by the Ohio Division of Wildlife, but was never protected by massive dikes. As a result, about seventy-five percent of the marsh was inundated by the storm of 1973 and has remained under the lake's influence: most of the time it is under water, but occasionally strong offshore winds uncover acres of mudflats.

Maumee Bay State Park

The former **Mallard Club Marsh**, 200 acres on Maumee Bay at the end of Cousino Road, borders the Cedar Point Marsh on the west and south. Beginning in 1974, it was acquired by the State of Ohio as part of the new 1,850-acre Maumee Bay State Park, which takes in about three miles of the shoreline. The marsh has been protected through the years by a large sandstone barrier. In 1992, the marsh and another 200 acres of old farmland were designated the Mallard Club Marsh State Wildlife Area. The Ohio Division of Wildlife plans to restore the entire area to marshland, managing its water levels and its wildlife in the same way as in Magee Marsh, but allowing open hunting as in Metzger Marsh.

Maumee Bay State Park is designed to serve all types of outdoor enthusiasts: birders, bathers, campers, hikers, boaters, and golfers. A hiking trail will pass through two miles of swamp forest and marshland in the wildlife area. A hill has been created for sledding in winter and hawk-watching during the spring and fall migrations. The bay beside the park is a favorite feeding and resting place for diving birds, which often gather by the thousands.

The marshes described above are owned or managed by state or federal agencies, which can afford the heavy costs of flood control. East of Turtle Creek, where sand ridges are higher, more of the marshlands remain in private ownership. The largest private marshes are Sand Beach Marsh and the Toussaint Marsh, but a number of small marshes along the Toussaint River are also privately owned.

Sand Beach Marsh

Hankison Marsh, better known as Sand Beach Marsh, originally covered 500 acres between Navarre Marsh and the subdivisions east of Locust Point. In 1922, 350 acres of the marsh were drained and subdivided, with a golf course and a bisecting concrete road connected to State Route 2. Like Pintail and Metzger Farms, this subdivision was flooded in the spring of 1929, and only the portion on the beach ridge was saved; the remainder reverted to marsh. According to former Magee Marsh manager Karl Bednarik, during aerial waterfowl counts in the 1950s the submerged concrete road could still occasionally be seen from the air. In 1931, a shooting club was organized, with five miles of dikes and five pumps for water control. In 1973, 100 acres along the south border were acquired for the Davis-Besse plant. The rest of the marsh is still operated as a private club.

Toussaint Marsh

Toussaint Marsh covers about 1,500 acres, bounded roughly by Rusha Creek, Lake Erie, the Toussaint River, and State Route 2. It is essentially an estuary of the Toussaint River and Rusha Creek. The marsh has been owned by the Toussaint Shooting Club for over 100 years, and has been diked since the late nineteenth century. Maps of the Toussaint Marsh drawn in 1950, 1970, and 1977 appear on pp. 102–103, illustrating how its plant communities changed in response to high lake levels.

The early days of the Toussaint Shooting Club are better known than those of most clubs, thanks to John G. Ketterer. In 1969, Ketterer wrote an account based on the records of David B. Day, a club member from 1905 to 1947.[20] According to early guides, the river was visited in the middle 1800s by hunters from Cleveland, who anchored their schooner there and used it as a base for hunting operations.

In February of 1878, the club was formed by a group of seven Cleveland sportsmen. It was not an easy place to reach by land, and some members were forced to travel a day and a half by train, buggy, or sleigh. Later, shooters from Cambridge, New Philadelphia, and Steubenville joined the club. Through the years membership has varied from twelve to twenty.

During the latter part of 1918, the U.S. Government installed the Erie Proving Ground for armaments testing and took over about 300 acres of Toussaint Club property, granting the club fin, fur, and feather rights. The clubhouse was moved from the Lake Erie shore, where it was in the line of fire, to its present location on the west side of Rusha Creek. Controls are maintained through six miles of dikes and two pumps.

Although it is commonly believed that hunters in early days always shot large numbers of ducks, Ketterer's history tells us that in 1893, the total kill of all twenty Toussaint Club members averaged only thirteen ducks per day during the season. This was due largely to the difficulty of travel through the marsh before the canals were dug. At times, great concentrations of ducks were seen, including 40,000 on October 13, 1939; 50–60,000 on October 28, 1939; and 50–60,000 on November 21, 1941.

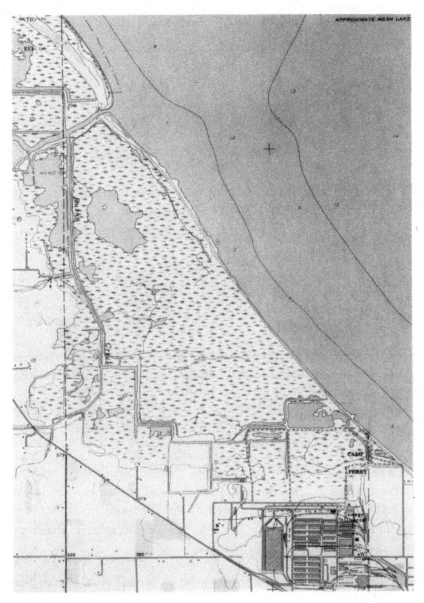

Toussaint Marsh, U.S. Geological Survey, 1967.

One of the most valuable sections of the Ketterer history is a list of game birds shot from 1893 through 1900. The following tabulation is the total for the eight years when there were twenty members:

Redhead	8	Bluebill (scaup)	172
Mallard	3512	Black Duck	891
Widgeon	496	Pintail	590
Shoveler	255	Gadwall	81
Wood Duck	242	Blue-winged Teal	608
Green-winged teal	600	Ruddy Duck	2
Canada Goose	4 (1897)	Bufflehead	13

The total is 7,470, with an average of 935 waterfowl per year, or 0.6 birds per acre of marsh. Note the scarcity of Canada geese.

The list of other birds shot is even more interesting, and indicates how hunting has changed:

Quail	400	Plover	181
Woodcock	200	Rails	110
Snipe	1044	Coot	301

All of these birds were considered less desirable than ducks, and once the marshes became more accessible, members concentrated on duck hunting. Note the extraordinary number of quail, which have now disappeared, probably because their preferred brushy habitat was destroyed through cultivation or development.

A number of small, privately owned marshes lie in the low-lying areas east of Turtle Creek. Some of these border on the larger marshes, others border the Creek or the Toussaint River. Most of these marshes are far enough from the lake that massive dikes are not necessary for their survival.

    1. Flaget Grodi Marsh, thirty acres on State Route 2 east of Turtle Creek.

    2. Witt Marsh, forty acres north of State Route 2 near the Davis-Besse Power Plant.

3. G. Ogden Trenchard Marsh, also known as the Toussaint Fur Farm, forty acres situated on the east side of the Toussaint River between Lake Erie and State Route 2.

4. Turtle Creek Gun Club, forty acres south of Route 2 and east of Turtle Creek.

5. Wild Acres or Gaeth Marsh, forty acres west of Route 2 along the Toussaint River.

6. Zetzer's Marsh, forty acres south of Route 2 along the Toussaint River.

7. The former Rockwell Club, now owned by T. Cornell, 302 acres in two sections between the Toussaint Marsh and State Route 2.

8. Camp Perry Marsh, 300 acres along Lake Erie in Erie Township, Ottawa County. Formerly owned by the Toussaint Shooting Club, it was acquired by the State of Ohio in 1918. The club retained fin, fur, and feather rights.

9. Sweeney's Marsh, fifty acres west of the Darby Marsh and north of Route 2.

10. M. Baumgardner Marsh, forty acres west of Darby Marsh and north of State Route 2.

11. Wesniak Marsh, part of the former Duck Haven Shooting Club. Sixty acres of wetlands south of Darby Marsh.

Chapter 3

Birds of the Marshes,

1870–1991

In country, suburb, or city, at any time and in any season, anyone of any age, whether active, sedentary, or bedridden, can enjoy watching birds. Many birders come to enjoy keeping records of their findings, especially when they can add a new species to the list. In any group of birders there will usually be at least one or two dedicated souls who carefully compile their own lists and those of friends into a permanent record, noting how often each species has appeared and in what numbers. These records show whether a given species is commonly or rarely seen, in large numbers or small; they also identify peak times for migration and the best places for birding.

I have been keeping track of bird records in the Toledo area for more than half a century. These statistics, which at first glance may seem to be merely dry lists of numbers, are actually full of stories: stories of a changing ecology, of successful adaptation, or of discouragement and failure. They tell of eagles struggling to raise their young in an age of DDT; of common terns stubbornly returning again and again to nest on the transitory mud islands in Maumee Bay; of the steady flap of great blue heron wings, decade after decade, bearing these majestic birds home to the West Sister Island heronry on a summer evening.

How trustworthy are these bird records from a scientific point of view? We can assume that the most accurate reports are those of easily identifiable species like orioles, goldfinches, and yellow warblers, or those with distinctive calls or songs. But what about species that confine themselves to dense cover, such as Connecticut and mourning warblers or certain sparrows? Are they really as rare as we think, or have they simply escaped detection? The opposite problem may occur on a bird trail, where the same individuals may be counted again and again as they move ahead of the birder. The best corrective for these kinds of errors is to assemble counts from a great many people over a great many years; as trends appear, individual errors become less important.

The most serious obstacle to accuracy is errors in estimating numbers, especially large numbers of water birds or migrating predators. Here the best estimates are made by observers in groups, where individual counting errors tend to cancel each other out. The National Audubon Society has institutionalized this approach in its annual Christmas Census, conducted each year all over the country between December 16 and January 3. The statistics in this chapter are based on my annual compilations of birders' reports, along with published records and my own observations. A complete list of birds in the western Lake Erie region appears in Appendix B, page 193–213.

Wetlands hold a powerful attraction for migrating and breeding birds, which take full advantage of the various habitats a marsh can offer. Each year, branches of the Mississippi and Atlantic flyways bring thousands and thousands of migrating waterfowl to the food and shelter of the Lake Erie marshes. When these and other migrating birds are halted by storm fronts or unfavorable winds, they pile up in the marshes in huge concentrations. Even in good weather, the marshes offer food and rest on the long journey.

In spring, northbound birds are reluctant to fly across the open waters of Lake Erie; they either resort to island-hopping from the Catawba Island sector to Point Pelee, Ontario, or they detour around

National Audubon Society Annual Christmas Bird Census, Ottawa National Wildlife Refuge, 1971–1988

This count is conducted in a circle, fifteen miles in diameter, that includes Magee Marsh Wildlife Area, Crane Creek Wildlife Area, Turtle Creek, Toussaint River and Metzger Marsh Wildlife Areas, Ottawa National Wildlife Refuge, Camp Sabroski, Navarre Marsh Wildlife Area, and a part of Oak Harbor. Counts are led by Tom Bartlett of Tiffin, Ohio and Cheryl and Ed Pierce of Akron.

The following species are birds observed on all eighteen counts:

Great Blue Heron	Great Horned Owl
Canada Goose	Downy Woodpecker
American Black Duck	Blue Jay
Mallard	European Starling
Bald Eagle	Northern Cardinal
Northern Harrier	American Tree Sparrow
Red-tailed Hawk	Song Sparrow
American Kestrel	Swamp Sparrow
Ring-necked Pheasant	Dark-eyed Junco
Herring Gull	Red-winged Blackbird
Rock Dove	American Goldfinch
Mourning Dove	House Sparrow

The following species are birds observed on at least 80% of the counts:

Snow Goose	Horned Lark
Common Goldeneye	Tufted Titmouse
Rough-legged Hawk	White-breasted Nuthatch
Great Black-backed Gull	White-crowned Sparrow
Northern Flicker	

One hundred twenty-two species have been observed in the Christmas Census, plus six additional species that have been seen during the count period but never on count day. The average is sixty species per year, with a low of forty-eight in 1980 and a high of ninety-one in 1984. The average number of individuals is 25,667 per year, with a low of 8,274 in 1980 and a high of 54,255 in 1984. The number of participating birders has ranged from five in 1980 to thirty-eight in 1986. —*Tom Bartlett* the west end of the lake at the mouth of the Maumee River. The Maumee Bay detour is the route favored by crows, hawks, waterfowl, snow buntings, horned larks, and other species. In fall, many of these birds recross the Detroit River and move southwest across Monroe County, Michigan and Lucas County, Ohio.

About 300 varieties of birds have been seen in the Lake Erie marshes. Of these, 167 nest or have been known to breed in or near the marshes. Some species that were once common have been eliminated, either by hunting or by radical habitat changes due to logging and farming. Among them are wild turkey, last noted in 1892; prairie chicken, 1875; ruffed grouse, 1905; passenger pigeon, 1885; and raven, 1890. The last breeding sandhill cranes were seen in 1880.[1] Although they no longer live in the marshes, all of these birds—with the exception of the passenger pigeon—are plentiful in other areas of the continent. Great egrets were virtually exterminated before 1915, but they began to reappear in the Lake Erie marshes in 1930, and West Sister Island now supports a healthy breeding population.

The confinement of the marshes, the clearing of the forests, and the cultivation of the cleared land profoundly affected the makeup of breeding and migrant bird populations—a phenomenon that has been investigated in detail by naturalist Harold Mayfield.[2] Several species increased their numbers dramatically. Some of these, like the cowbird, horned lark, bobolink, savannah sparrow, and vesper sparrow, were birds from the short grasslands of the western or northern prairies; others, like the barn swallow, purple martin, chimney swift, and phoebe, took up residence in man-made structures. Among those that thrived in the cleared spaces of lawns, fields, and orchards were mourning doves, robins, eastern and western meadowlarks, red-winged blackbirds, grackles, and cardinals.[3]

Crows replaced ravens, but after peaking in the 1930s they have since become rare breeders near Lake Erie. Away from the lake, they have increased. Northern bobwhite (quail), once common, no longer appear in or near the marshes. Ring-necked pheasants, introduced early in the century, reached a high point in the 1940s and are now in serious decline, as are bobolinks and grasshopper sparrows.

A careful study of bird population trends in Sandusky's Winous Point Marsh was published by John Anderson in 1960.[4] He listed all the species seen during the first week of July in 1880, 1930, and 1960—a total of 118 species, with a one-year maximum of eighty, in 1960. Somewhat surprisingly, the overall change was comparatively minor: eighteen species were more abundant in 1960 and eleven were less. Another surprise was an increase in breeding ducks.

Within the last half-century, resident predators near Lake Erie have undergone interesting transitions, some of which are difficult to explain. Barn owls, red-shouldered and Cooper's hawks, and harriers have either become rare breeders or have stopped breeding altogether. Eagles fell to a very low level, then recovered. On the other hand, great horned owls and red-tailed hawks have increased, while screech owls have remained fairly constant. Two species with comparable feeding habits have reacted quite differently: the loggerhead shrike has almost disappeared while the kestrel, or sparrow hawk, has increased, apparently because kestrels are comfortable with people.

The spring migration brings a great many hawks to the Lake Erie marshes. Joe Komorowski has been recording their numbers and species for at least five years. In 1991, Mark and Julie Shieldcastle joined the project, which was then expanded to include a number of other volunteers. On forty-two days between February 21 and May 10, 1991, 3,951 migrating hawks and owls were counted, and 146 hawks were banded. Records for any given day include flight and wind directions, wind speed, sky condition, and the time of day the birds were seen. The organizers hope to expand the study still further in future years.

Bald Eagles

The most spectacular—and most publicized—bird living in and near the marshes is the bald eagle, famous for its rarity, great size, and immense nest. Vermilion, Ohio holds the record for the biggest eagle's nest ever recorded: twelve feet high, eight and a half feet across. Eagle nests have been found near the marshes in Lucas, Ottawa, Sandusky, and Erie Counties, and on three of the Lake Erie Islands: West Sister, Green, and Kelley's. One nest has been seen inland, in Wyandot County. The birds feed on small mammals as well as fish, which the inland nesters apparently obtain from nearby creeks and canals.

Eagle records from 1921 to 1934 are limited to my personal observations, since birders were fewer in those days. At that time there were five nests in Lucas County: one was near Maumee Bay, another just south of the Cedar Point Marsh, and one each on West Sister Island, the Cedar Point Marsh, and the Ottawa Refuge. There were four nests between the cities of Toledo and Monroe, Michigan, one of which—on North Cape—was occupied continuously from 1934 through 1953. There are no nests in that area now.

Bird bander, naturalist, and former game protector Laurel Van Camp published the first complete survey of Toledo area eagle nests in 1934 and again in 1974.[5] In 1934, Van Camp listed three nests in Lucas County, eight in Ottawa, three in Sandusky, and two in Erie—a total of sixteen nests. Unfortunately, two of the Lucas County nests had already been destroyed. In the later survey, covering the years 1959–74, peak numbers of nests in the four counties were fifteen in 1959, eleven in 1962 and 1963, and thirteen in 1966. After that, numbers fell rapidly to only four nests in 1977.

The number of nests in which young were reared also plummeted after the late 1960s. While the success rate for 1959 through 1968 had averaged thirty percent, in the next decade it fell to half that number. No fledglings at all were raised in 1963, 1971, 1972, and perhaps in the following few years, when records were few. The reason, biologists discovered, was that the female eagles were eating fish containing mercury, pesticides, and chemical fertilizers. These contaminants prevented egg shells from hardening, and the eggs collapsed as they were laid.

Since the birds continued to breed year after year, the Ohio Department of Natural Resources began a program to supply nesting eagles with eggs or young from the Cincinnati and Columbus Zoos, the Cleveland Museum of Natural History, and the U.S. Fish and Wildlife Service's Patuxent Research Center. The fledglings were readily accepted by the wild birds, and the practice continued until 1987, when it was clear that the eagles' natural fertility was returning. In 1982, eight young were raised in five out of the seven area nests, three of them in a single nest. After that, active nests rose to eleven in 1988, twelve in 1989, and sixteen in 1990. In fact, two additional nests were built in the fall of 1989 for use in 1990, one near Port Clinton and the other near Gibsonburg. This remarkable comeback was helped along by the ban on DDT as well as by the fostering program and a policy of protecting active nests. Wildlife biologists estimate that if the eagles can maintain an average of seven young for every eleven nests (sixty-four percent), they will be able to maintain the population. In 1990, the sixteen nests produced twelve eaglets (seventy-five percent).

Breeding eagles also face a housing problem. In recent decades, as trees have been destroyed by Dutch elm disease, lumbering, and development, eagles have been forced to build their nests in cottonwoods, few of which are large or strong enough to bear the weight. On February 15, 1967, no fewer than four nest-bearing trees blew down in a storm. Other nests were lost in 1981, 1982, and 1988, killing several nestlings. Sometimes the birds build two nests in the same area, apparently to meet emergencies.

In earlier days, when nests six feet high and about eight feet across were built in sturdy elm trees, they would be used year after year. Renovated annually with a few sticks and a new lining, a nest would remain in service as long as it held together and the surrounding woodlot was not leveled. Rarely, a nest would be visited by too many people, usually birders and picnickers. For example, in 1961 a nest was built beside a parking lot in the Crane Creek State Park near the Lake Erie shore. When the fledglings arrived, scores of observers and photographers turned up. If the female refused to stand up on the nest for a picture, someone would pound the tree with a club. As might be expected, that nest was abandoned.

However, eagles did not always avoid human company. The nest holding the area record for longest continuous use was built in a large elm in the Cedar Point Marsh. It was occupied from at least 1920 until 1945, and for much of that time it stood only a few hundred feet from a unit of the Lamb commercial fishery, which worked from a two-story house and several net sheds on the lakeshore. The birds may have fed on the discarded fish.

Perhaps the oddest location for an eagle's nest was discovered in Sandusky County near Muddy Creek on March 31, 1976.[6] The nest, which was of normal diameter, had been built on the ground in a field planted with soybeans the previous year; it was made of bean straw piled up to form a mound, and it held two eggs. Unfortunately, within a few days both eggs disappeared and the nest was abandoned. A tower designed by federal and state employees was built by the Toledo Edison Company in the hope that the eagles would nest on it, but they were not interested. However, the idea appears sound, since similar structures are used by ospreys. Perhaps the eagles will begin to see the advantages of nesting platforms as large trees become more and more scarce.

Van Camp's 1974 report on area eagles describes another unusual event, in 1960, when two young eagles apparently fell out of a nest near Sandusky. The landowner built a large pen out of snow fence to keep the birds from wandering away from the nest site. The effort paid off: the penned birds survived and flew away, though a third young eagle still in the nest was found dead a few days later.

Breeding eagles have a curious habit of nesting in great blue heronries, a practice that has occurred four times in the general area of the Lake Erie marshes. The first was in 1937 at the mouth of the Ottawa River at North Cape, Michigan, in a wooded area that has since washed away; the second was on West Sister Island in 1959; the third in 1965, in a very large heronry in the Winous Point Marsh on Sandusky Bay; and finally, in 1987, in a heronry south of Route 2 near the Sandusky Bay bridge.

It appears that bald eagles will be residents of the Lake Erie marshes for years to come, in addition to those that appear during the spring and fall migrations. Concentrations of eagles are rare, however. On March 20, 1937, I saw seven adults in a tree west of the Cedar Point Marsh, and on March 31, 1948, Laurel Van Camp counted fourteen feeding on offal near an Oak Harbor slaughterhouse. After that no large groups were seen until September 4, 1988, when thirteen—including only one adult—were spotted in the Ottawa Refuge. By September of 1990, the number of area eagles had grown to eighteen, including two adults.

Herons

Because of their size, statuesque bearing, and tolerance of people, herons as a group are an important feature of the marshes, adding life and beauty to the summer wetlands. No fewer than twelve species of herons can be found here, although six of the species are uncommon and two of those—the American and least bitterns—are so inconspicuous that they are rarely seen.

Herons nest in three locations that are not part of the Lake Erie marshes covered in this book: West Sister Island, the Winous Point Marsh, and an area near the Sandusky Bay Bridge. The most important is West Sister Island, an eighty-two-acre, wooded tower of limestone rising up in Lake Erie about nine miles from the mainland.

When water levels are low in portions of a marsh, food for herons becomes concentrated in those areas and may attract masses of feeding birds. The largest area heron count on record was made on June 29, 1976 by birder Tom Kemp, who estimated 1,000 great blue herons, 400 great egrets, and an equal number of black-crowned night herons. On June 12, 1988, Joe Komorowski counted 1,078 great blue herons in the Cedar Point Marsh. Ordinarily herons are scarce in winter, but an amazing 327 great blues were found in the Ottawa Refuge on December 26, 1976.

The **great blue heron,** four feet tall with a wingspread of about six feet, is the largest breeding bird in the area. It can be found any time of year, wherever it can find open water in winter. Many breeding pairs gather together in a suitable wooded area to form a heronry, sometimes miles from the nearest wetlands. Because the food supply is far from the heronry, there is no competition, and nests can be built very close together. About three young are hatched per nest.

Strewn with discarded feathers and whitewashed with droppings, these colonies are well documented because they are so noticeable.

The first one I saw was a group of twenty-five nests at the outer edge of Magee Marsh in 1929. E. L. Moseley reported 1,118 nests north of Hessville in Sandusky County in 1935,[7] but they were later destroyed by logging. From 1935 to 1941, thirty-two pairs nested in a woodlot on North Cape next to the Erie Marsh. No one knows when the Winous Point colony was established, but it is still in use today. The colony flourished in the 1960s and 1970s, with a peak of 1,600 nests in 1973, but high water levels in recent years have reduced the number to 1,000.

Although it had been known since the turn of the century that herons nested on West Sister Island, it was not until 1945 that a complete breeding survey was conducted on this relatively inaccessible spot. The count revealed some 400 heron nests, which increased to 450 in 1959 and 600 in 1976. That number remained constant until 1987, when it rose to 1,400. Back on the mainland, the most recent heronry was built in 1978 south of State Route 2 near Sandusky Bay. Its 100 nests are visible from the road.

The second-largest heron, the **great egret,** has a far more dramatic history. As early as 1870, Emery Potter wrote, "Battalions line the beach [of Maumee Bay] like so many ghosts in white."[8] Jerome Trombley commented that occasionally they were abundant in July and August: "[It] undoubtedly breeds, but where?"[9] Charles F. Elliott, whose family was for many years in charge of the lighthouse on West Sister Island, told me in 1967 that when he was a boy, at the turn of the century, both great egrets and great blue herons nested on West Sister.

Egrets declined rapidly after 1900, due no doubt to the slaughter of the birds for fashion feathers. In 1912, Trombley said, "The keeper of the Monroe Marsh [Michigan] has not seen one in ten years or more."[10] Beginning in 1926, my own records show that a few were seen in the marshes each year until the drought years of the early 1930s brought an increase: forty-six were seen on August 9, 1930, and by 1933 there were 215 in the general Toledo area.

Like the great blue heron, egrets nest in colonies. The first nests were reported by Lawrence Hicks on Eagle Island near the Winous Point Marsh in 1940 and 1941.[11] In 1945 nests were found, but not counted, on West Sister Island. By 1959, 125 had been counted on the island, rising to 600 by 1976 and approaching 1,000 in 1991. Breeders at Winous Point reached a peak of twenty-five pairs, according to John Anderson, then dropped in 1960 to a single pair before vanishing altogether. Great egrets have been found during the winters of 1981, 1987, and 1988.

Whether it was because the **black-crowned night heron** was smaller and inconspicuous, feeding mostly at night, or because observers were few and had little opportunity to check the marshes, or because the birds actually were limited in numbers before 1900, we will never know, but all of the early bird authorities were in agreement that this species was seldom seen. Emery Potter did not mention it. Jerome Trombley, who lived near the Erie Marsh, said that it was "nowhere common in Michigan, found singly and at long intervals."[12] J. M. Wheaton (1879) wrote, "I do not know positively of its breeding within the State of Ohio."[13] In 1903, in his *Birds of Ohio*, Lynds Jones wrote: "Locally common in the state but seems to be absent from many regions. Migration dates are lacking."[14] In spite of all this, in 1930 my first area checklist described it as a common summer resident.

Early night heron colonies included one or two in the Cedar Point Marsh before 1920; one colony of 20 nests in Magee Marsh in 1928; a small but conspicuous group at the border of the Toledo Beach summer resort not far from the Erie Marsh, during the 1920s; and about 175 nests at North Cape, just over the Michigan line near the Erie Marsh, in 1938 and for several years thereafter. The night heron colony on West Sister Island was first mentioned in 1934 by Fred Brint, a federal game protector, who estimated that it consisted of 5,000 to 10,000 nests. However, Brint may have overestimated the number. It is also possible that these herons had been present on the island in the early 1900s but had been overlooked. The first count by area birders of black-crowned night herons on West Sister was in 1942, when 1,000 nests were estimated. This figure increased in 1959 to 1,200 and again in 1983 to 1,500. Numbers dropped to 1200 in 1991.

The **green-backed heron** is much smaller than any of the previously mentioned birds—a little over one foot high. Its habits are distinctly different from those of other herons, since although it commonly roosts in trees, it is a loner in both nesting and feeding.

It is also more adaptable than other herons and can be found wherever there is water: in marshes, ponds, creeks, even large drainage ditches. Its numbers have remained very constant through the years.

American and least bitterns differ greatly in size. The least, as might be expected, is the smallest of the local herons, less than a foot high and very slender. Its contrasting colors of buff, chestnut, and black make it the most colorful of the herons. Both species are solitary breeders, preferring to nest in tall grasses or cattails. The male American bittern is famous for his marvelous call at mating time. For some unknown reason, the bitterns have become very rare, even though their marshland habitat—with the possible exception of wet prairies—has remained stable. In the past, both species probably have been regular summer residents, and the American bittern has been found several times in winter.

Seven species of heron have introduced themselves since 1920; all, with the exception of the cattle egret, were rare during their first years. The **snowy egret** was first reported in 1934 and has been seen fairly often since 1967, but is never numerous. It nested for the first time on West Sister Island in 1981. The **little blue heron** was first seen in 1924, but not with any regularity until 1965, and was not found breeding until 1982, on West Sister Island. Unlike other species, immature little blue herons are white. The largest number seen on one day was thirty-five, in 1939. The **tricolored heron** was first listed in 1965, and one has been seen about every year since, except 1976, 1977, and 1984. The greatest number seen at one time was four, in 1981.

The **cattle egret,** now seen regularly in the Lake Erie marshes, migrated from Africa to South America in the nineteenth century. The first one reported in the United States was seen in Massachusetts in 1952; since that time it has been seen in varying numbers in every state except Alaska. The first noted locally was at Magee Marsh in 1960. By 1978, there were twenty cattle egret nests on West Sister Island, but they later moved to Turning Point on Sandusky Bay. The average annual number reported for the past thirteen years is ninety-five, but 254 individuals were reported in 1980. Only one winter sighting is on record.

The **yellow-crowned night heron,** a southern Ohio bird, was first discovered locally in 1940 and has since been seen approximately every two years. A pair raised two young along the Raisin River east of Monroe, Michigan in August of 1983. The bird has been seen during summers of previous years and may have nested in other locations as well.

The **glossy ibis** was first reported in 1943 and has been noted about every other year. An amazing twenty-six were seen between April 21 and May 23, 1962. The **white-faced ibis,** which closely resembles the glossy ibis, was definitely identified in 1984 and 1985 and may have been mistaken for the other species in some earlier records.

Based on reports I have received in the last seven years, American bitterns are down close to fifty percent, least bitterns are down thirty-five percent, the snowy egret has doubled, the little blue heron is up thirty-seven percent, and the cattle egret is down twenty percent. No glossy ibis or yellow-crowned night herons were reported in 1987 or 1988, but yellow-crowned night herons were noted in 1990 and glossy ibis in both 1989 and 1990.

Cormorants

One of the most interesting developments of the 1980s was the rapid increase in the number of double-crested cormorants. From 1925 until 1970 the number of cormorants seen locally per year ranged from zero to a high of fifteen in 1946. The largest area count was sixty-four on October 21, 1943, at South Bass Island. On fishing trips during the 1940s and early 1950s, I noted a dozen or so pairs nesting on one of the Chick Islands in Canadian Lake Erie.

Local cormorant populations remained low except at South Bass Island until 1981, when thirty-one were seen at Maumee Bay Shore, seventeen on the Toussaint River, and 190 near the Lake Erie islands. After that the birds began to gather at two locations: Middle Harbor on Catawba Island and, to a much lesser degree, the Ottawa National Wildlife Refuge. In 1982, 240 were seen at Middle Harbor, and flocks have been recorded every year since, with a high of 1,100 on November 7, 1983. Beginning in 1981 and through 1984, from two to six were found in the Refuge, and during the following three years they were found each month from June through October in gradually increasing numbers—a maximum of 42 in 1985, 39 in 1986, and 113 in 1987.

In 1987 six pairs of cormorants built nests in dead trees on the Refuge near the mouth of Crane Creek. Unfortunately, five of the six nests were destroyed by a windstorm on August 2, and the sixth appeared to have been abandoned. Whether there were eggs or young in the nests remains unknown, but it was clear that young cormorants were being raised somewhere not too far away, perhaps in Michigan: on September 13, eight first-year birds were seen in a flock of forty-seven, in the same area of the Ottawa Refuge where the nests had been built earlier in the year.

As an indication of total numbers in the general area, a flock between Middle and North Bass islands was estimated at 2,000 on September 25, 1987, and naturalists from the Ohio Division of Wildlife on an aerial waterfowl survey saw 2,000 in Sandusky Bay on November 16, 1987. One thousand were counted in the Ottawa Refuge on November 4, 1990 and 660 near Marblehead on November 11. A surprising total of ninety—considering the season—were seen in Ottawa on August 4, 1991 and 2,298 were seen in Cedar Point Marsh on October 13 of that year. But the most astonishing change took place in 1992, when an estimated 180 active nests were found on West Sister Island. The last known nesting in Ohio by these birds had been in the 1880s, at Grand Lake St. Mary's in Mercer County.

Gulls and Terns

Ring-billed and herring gulls and common terns form an important segment of marsh bird life, even though there are no breeding colonies within the wetlands themselves. These birds prefer to nest on bare, isolated stretches of sand or mud in or near Maumee Bay. Until 1940, **ring-billed gulls** were common, but not as numerous as herring gulls. Since that time, they have increased to the point that herring gulls now outnumber them only in winter, after the late-December exodus. Through the years, state wildlife biologists have banded 18,000 juveniles, of which one hundred were recovered.

On July 17, 1966, Ray Frankhouse and I found the first breeding ring-bills in Ohio, on one of the mud islands formed by dredging in Maumee Bay. The following year there were fifty nests with eggs or young. The ring-bills bred through 1970, but by the next spring, wave action in the bay had worn the man-made islands too low for nesting. In 1975, the Lucas County Port Authority built a dredge waste containment area at the mouth of the Maumee River. Surrounded by three-and-a-half-miles of high, rocky dikes, this area proved to be an ideal nesting spot for ring-bills, enhanced by an abundant supply of small gizzard shad and other food fish from the warm water outlet of the nearby Toledo Edison Bay Shore power plant. The few pairs that bred in 1975 increased to two hundred pairs in 1977, then exploded to six thousand in 1981.

Ring-bills continued to nest successfully until the intense heat wave of June and July, 1988, after which I found the skeletons of about 1,500 fledglings. All the eggs were gone, and only about 150 young ring-bills—those that had developed quickly enough to avoid the heat—were flying about over Maumee Bay. Adults had decreased from 6,000 to a mere 750; since they must have survived the heat, they had apparently left the area. In 1989 and 1990, nest numbers were back to normal, and at least 3,500 young were raised successfully. But in 1991, although many eggs were laid, no fledglings survived, and in 1992 there were no nests at all.

Ring-billed gulls have become noticeably more accustomed to people in recent years. They walk about on the Metzger Marsh fishing pier, searching for discarded minnows, even when anglers are about. At Crane Creek State Park, flocks of more than fifty roam around the picnic grounds looking for scraps. On December 8 and 9, 1990, Paula Jack and her associates estimated 750,000 ring-bills in the Marblehead district east of Darby Marsh, by far the largest number ever reported in sixty-four years. The former record number was 45,000 on December 14, 1986. The 1990 number is especially interesting in view of Dr. Lynds Jones's comment in 1903: "I have looked for this gull in vain."[15]

Except during the breeding season, **herring gulls** have always been numerous. As early as 1939, 3,000 were reported on the local Christmas Census, and 10,950 on the census of 1980. Flocks of 3,000 or more are not uncommon in late December or early spring, when they travel miles from Lake Erie in search of food at shopping centers or newly ploughed fields. They are least common from late December or early January to the second week of February. Winter birds may come from great distances; one banded bird found dead in this area had been hatched on Kent's Island, New Brunswick, and an adult banded here was later captured at West Dover, Nova Scotia.

The first local herring gull nest was found in 1953 by Lawrence Shafer, on an island dredged from the Toledo ship channel in Maumee Bay; a second was found in 1954. By 1965, a small group had settled on West Sister Island and has remained there ever since, reaching 163 nests in 1976 and 600 in 1987. The Maumee Bay group increased to ninety pairs by 1967. When the disposal unit was built, herring gulls moved in along with the ring-bills, and since 1977 have averaged about 250 nests per year. The two species often nest very close to each other.

A **laughing gull** visited the ring-bill colony in the disposal unit every year from 1978 through 1986, except 1982. In May of 1984, I found a female sharing a nest with a female ring-bill. The nest contained two laughing gull eggs. On June 1, one ring-bill egg had been added, and the two females took turns incubating the eggs. Unfortunately, on June 9 I found the nest empty, and the laughing gull had disappeared.

Bonaparte gulls do not nest near the marshes, and for that reason I have not included them. For the record, however, a new high of 250,000 individuals was reported on December 8 and 9, 1990 by Paula Jack's group.

The story of the attempts of **common terns** to breed in the Toledo area, revealed through my yearly records, reads like a fable on the virtue of perseverance. Their first breeding attempt was in 1928, when a group of fifty tried to nest in Metzger Marsh, but failed because of changing water levels. The group then moved to the Little Cedar Point sandbar, where it increased to 500. In spite of the delayed season, fifty-three young were fledged that year, and from 1928 to 1934, over 5,000 fledglings were hatched on the Point. In 1932, 1933, 1935, and 1936, many eggs were laid only to disappear; in those depression days, it was rumored that people gathered them for food.

In 1937, the terns moved to a dredged island in Maumee Bay and raised one thousand chicks. In 1939, they moved back to the Point and failed again—why, I couldn't determine. No terns at all bred from 1940 through 1951. At that point, new islands were piled up beside the ship channel all the way to Harbor Light, and from 1952 through 1955 close to three thousand terns bred successfully. Then the Lake Erie water level rose again and these islands, too, were washed away. Once more new islands were created in the Bay, and from 1965 through 1967 they were occupied by five thousand terns, but as the mud succumbed to wave action yet again, fledglings fell to eighty. For the next eight years, there were no nesting areas available to them in the Bay.

In 1975, the terns made use of the newly built disposal unit at the mouth of the Maumee River and, despite the crashing sounds of tractors, raised over four hundred young. Numbers held up through 1980, when 147 were banded, and then collapsed—apparently because the terns could not compete with the ring-billed gulls. In 1981, six hundred attempted to nest with no results, and in following years, the number of nests dropped from forty to zero. There is no shortage of migrating terns in the area, however. Their peak number was reached in August and September 1976, when flooding in the Ottawa Refuge created two square miles of mudflats between Veler and Ottawa-Lucas roads, attracting seven thousand common terns and three thousand Forster's terns.

In 1987, biologists from the Ottawa National Wildlife Refuge attempted to induce common terns to breed on an enlarged dike on the Lake Erie shore near the mouth of Crane Creek. They had little success: two nests with eggs were found, but the eggs failed to hatch. However, in 1989 a small group did breed successfully, so the tern story is not yet finished and may yet have a happy ending—until the next disaster.

Forster's tern was first discovered in the Toledo area on September 13, 1932. It has been seen regularly ever since, including many times during summer in recent years, but no evidence of nesting has been found. This species is much easier to identify in fall plumage. Peak numbers were 3,000 between September 3 and 19, 1976, and 3,125 on September 3, 1982, both in the Ottawa Refuge. No doubt Forster's tern will breed here in the future, since more than a dozen colonies now nest at the southern edge of Lake Huron and at the northeast end of Lake St. Clair.

Waterfowl

Because the first reason for preserving the marshes was hunting, waterfowl production has always been very important. But even before the local wetlands were disturbed, either comparatively few ducks bred here or few surveys were made. According to the earliest records, by Emery Potter and Jerome Trombley, nesting waterfowl were limited to a few mallards, black ducks, and blue-winged teal. Wood ducks, which formerly had been fairly common breeders, were at that time already badly reduced. John Anderson[16] lists as breeding populations mallard and blue-winged teal in 1880; mallard, black duck, and blue-winged teal in 1930; and mallard, black duck, pintail, green-winged teal, blue-winged teal, wigeon, northern shoveler, wood duck, and redhead in 1960. In 1940, I estimated the breeding duck population of Lucas County at one hundred pairs of mallards, two hundred pairs of black ducks, seventy-five pairs of blue-winged teal, a few pintails, and an indefinite but rather small number of wood ducks. As wetlands away from the marshes were destroyed, nesting ducks became concentrated in the remaining areas.

A completely different picture of waterfowl production was given by David B. Day in 1934.[17] At the twentieth North American Game Conference, Day asserted that ten to fifteen thousand ducks were hatched annually in a 1,500-acre Lake Erie marsh, presumably the Toussaint Shooting Club, of which he was a member. He apparently was hoping to pursuade the federal government to relax hunting restrictions. Unfortunately, his unsubstantiated statement

Aerial Counts of Waterfowl in and near the Lake Erie Marshes, 1985–1990

Aerial waterfowl counts are much more accurate than those conducted from dikes or boats—except for wood ducks, shovelers and blue-winged teal, whose preference for heavy cover makes them difficult to spot from the air. In his many aerial surveys from 1953 to 1957, Delmar Handley of the Ohio Division of Wildlife found that about 3,600 ducks per year were raised in Lake Erie marshlands. Since 1985, the goal has been to count total numbers of all waterfowl species. The counts are conducted from helicopters around the 1st and 15th of each month from August to January.

	Maximum number	*Date Seen*
Wood Duck	2,960	October 16, 1985
Green-winged Teal	10,330	October 13, 1989
American Black Duck	63,460	December 1, 1988
Mallard	142,050	November 14, 1980
Northern Pintail	14,475	October 15, 1986
Blue-winged Teal	2,425	September 6, 1985
Gadwall	12,425	October 29, 1985
American Wigeon	23,680	October 11, 1988
Ring-necked Duck	410	November 14, 1988
Canvasback (*)	10,000	January 3, 1990
Redhead	2,500	January 2, 1990
Lesser Scaup	46,122	April 16, 1990
Common Goldeneye	12,000	January 2, 1990
Common & Red Merganser	55,580	November 18, 1985
Ruddy Duck (**)	5,250	October 13, 1989
Total Ducks	210,385	November 14, 1988
Canada Goose	20,830	November 16, 1987
Snow Goose	182	November 1, 1989
Blue Goose	150	September 15, 1988
Tundra Swan	3,200	December 12, 1985
American Coot	14,315	October 15, 1986

(*) The largest recorded number, 15,000, was seen from the ground on February 3, 1985.

(**) The largest recorded number is 24,660, seen November 30, 1962.

—*Jack Weeks*

received wide distribution, even by the National Audubon Society, and resulted in serious overestimation of the number of breeding ducks in this area.

In 1951, the Ohio Division of Wildlife purchased Magee Marsh and began a program to encourage waterfowl production in the Lake Erie marshes, directed by Karl Bednarik. From 1953 through 1969, breeding mallards, blue-winged teal, and wood ducks increased greatly, while black ducks improved and then dropped back to their original numbers. Redheads reached a high of eleven nests in 1963 before decreasing again. Unusual breeders, including green-winged teal, gadwall, wigeon, shoveler, pintail, ruddy duck, and hooded merganser, peaked at four nests per year, mostly in the second half of the period.

The increase in breeding pairs was attributed to manipulation of water levels, natural cover, and food plants, plus the use of artificial nesting boxes, which proved very attractive to wood ducks. Nesting in these boxes rose from seventeen percent in 1964 to forty-two percent in 1973, and was probably the most important factor in the restoration of wood ducks to their present status. A few pairs of hooded mergansers also nested in the boxes.

At the end of the project, it was decided that future emphasis should be placed more on furnishing an attractive stopping place for spring and fall migrants than on increasing the comparatively few pairs of breeding waterfowl. The 1991 list of breeding pairs includes 150 wood ducks, 150 mallards, twenty-five black ducks, fifty blue-winged teal, five pintails, five shovellers, fifteen hooded mergansers, one ruddy duck, and three redheads.

The greatest change in the history of waterfowl populations in the Lake Erie marshes began in 1965 with the introduction of resident Canada geese at the Crane Creek Wildlife Experiment Station in Magee Marsh. In 1967, fifty pairs of Giant Canada Geese (*Branta canadensis maxima*), wing-clipped and weighing from ten to twenty-one pounds each, were obtained from the Mosquito Creek Waterfowl Refuge in Trumbull County, Ohio. This captive subspecies is more than twice as heavy as the two main species of migrant Canada Geese, the Common Canada Goose (*Branta canadensis canadensis*) from east of Hudson Bay and the Mississippi Valley Goose (*Branta canadensis interior*) from west of Hudson Bay.

The project was intended to last from 1965 through 1970, but proved so successful that it was extended. Breeding geese nested on the dikes as well as in metal tubs cut from oil drums and set in the marshes on wooden poles to foil egg-loving raccoons. It was found that goslings hatched in artificial nest holders tended to select nest holders when they reached breeding age three years later.

The gosling hatch was impressive from the first year, rising from 118 in 1967 to 1,105 in 1973 as the breeding flock increased. Numbers leveled off as the available area was occupied, and averaged 1,357 between 1974 and 1989. Migrating geese began to appear in growing numbers, attracted by the resident geese and the increased food supply. Before 1966, the maximum number seen in the entire Toledo area was 2,200 on March 24, 1946. In October of 1966, an aerial count of the Lake Erie Marshes and part of Sandusky Bay revealed 5,900 geese, a number that climbed steadily to the all-time high of 38,000 seen on December 4, 1977 in the Ottawa Refuge and Magee Marsh.

Reproduction data are based on birds observed in the annual July goose roundup in Magee and the Ottawa Refuge. An additional five hundred birds may nest in surrounding areas, and all resident geese are now free-flying. Although goslings are banded each fall, no attempt is made to confine, wing-clip, or restrain them as they join the resident flock. The sight of a pair of honkers with a fleet of young is now common in Magee Marsh, Navarre Marsh, and the Ottawa Refuge, and is spreading throughout all the marshes. Normally goose populations are confined to residents up to the end of September, when they are joined by migrating birds. The population reaches a peak late in October or in November. The size of the wintering population depends largely on weather and food conditions, and ranges from ten to twenty thousand birds.

Most of the geese taken by hunters are shot during the last week of October. In the thirteen seasons from 1951 through 1963 at Magee Marsh, 16,550 hunters took only ninety-eight geese; in 1981 through 1987, within the five-mile mandatory reporting zone, 12,300 hunters killed 1,750 geese per season, with a high of 2,154 in 1985 and a low of 1,399 in 1986.

Shorebirds

Locating and identifying shorebirds is so difficult that they are usually the last group the average birder becomes acquainted with. Several species change their appearance radically between spring and autumn, while others look so much alike that distinguishing between them is a major problem. Only five species—at one time there were six—nest in this area. Those that nest near the marshes have included the killdeer, spotted sandpiper, upland sandpiper, common snipe, woodcock, and Wilson's phalarope. Only one Wilson's nest has ever been recorded in Ohio; it was found on June 4, 1980 by Mark Shieldcastle. Piping plovers raised families on the Little Cedar Point sandbar until 1942.

The great majority of shorebirds are transients, journeying in spring from the Argentine to breeding grounds in northwestern Canada and Alaska, then back again in the fall. The earliest arrivals in spring appear between the middle and the end of February: killdeer, woodcock, and common snipe. These are followed in March by golden plover, pectoral sandpiper, and greater and lesser yellowlegs. Remaining species drift in during April and early May, and a few linger into the first week of June: least and semi-palmated sandpipers, red knot, and dunlin. Occasionally, a few unexpected individuals show up later in June, having somehow missed their trip to the Arctic. The first of July marks the beginning of the the great fall return migration. Normally, the first to make an appearance are short-billed dowitchers, stilt sandpipers, least and semi-palmated sandpipers, ruddy turnstones, greater and lesser yellowlegs, solitary sandpipers, and semi-palmated and black-bellied plovers.

Most shorebirds remain through October, their departure dates largely governed by the availability of feeding areas. A surprisingly large number of species have been seen occasionally in winter: black-bellied plover, greater and lesser yellowlegs, ruddy turnstone, least sandpiper, dunlin, woodcock, common snipe, purple sandpiper, Baird's sandpiper, red and red-necked phalaropes. One killdeer is usually seen every winter.

The local birder's biggest problem with shorebirds is locating mudflats where they congregate. A few species, such as piping plover, sanderling, and ruddy turnstone, prefer sand beaches; however, high water and large dikes have eliminated many western Lake Erie beaches. In spring, flooded farmlands are among the best places to see shorebirds. In autumn, they may assemble in portions of the Ottawa Refuge or Magee Marsh that have been allowed to drain. In undiked Metzger Marsh Bay, a strong southwest wind sometimes blows the waters offshore, creating acres of mudflats.

A total of thirty-one species regularly pass through the marsh area, plus five—red and rednecked phalaropes, ruff, purple sandpiper and piping plover—which are not reported every year. With the help of Mike Bolton, John Szanto keeps track of shorebird populations, sending his findings each year to International Shorebird Surveys at the Manomet Bird Observatory in Massachusetts. In 1991, on forty-one trips to Magee and Metzger Marshes, the Ottawa Refuge, and the Erie Marsh on the Ohio-Michigan line, twenty-eight species were recorded, including Wilson's and rednecked phalaropes. The total number of individual birds counted was 37,190, most of them dunlins. Most welcome to birders were two piping plovers.

From 1975 through 1990, the average number of shorebird species reported each season was thirty-three, with a low of thirty in 1975 and a high of thirty-seven in 1985. Shorebirds new to the marshes included a black-necked stilt seen in Magee Marsh by many observers in May, 1981 and from May to August, 1989; four long-billed curlews seen in the Cedar Point Marsh May 22, 1985; individual curlew sandpipers found in a field near the Ottawa Refuge (1985), in Winous Point Marsh (1987), and in Metzger Marsh (1989); and a sharp-tailed sandpiper seen by Szanto and Bolton on December 1 and 2, 1990, again in Metzger Marsh. This last was only the second time a sharp-tailed sandpiper has been seen in Ohio. The first recorded European whimbrel in Ohio was identified by Matt Anderson on the Maumee River rapids in July, 1988.

In the springs of 1973–75, Lake Erie flooded much of the farmland adjoining the marshes and demonstrated the great attraction of suitable mudflats to migrating shorebirds. The flooding was especially severe in the Ottawa Refuge, where cultivated land between Veler and Ottawa-Lucas Roads became a shallow lake. From 1973 through 1978, as the water receded after each spring peak, it left two square miles of ideal shorebird habitat. Actually, there were two groups that benefited from this situation: shorebirds and birdwatchers.

Many new species appeared in the area in unprecedented numbers. Through the six years, members of the Toledo Naturalists' Association counted over thirty species, some in surprising numbers: ninety Hudsonian godwits, sixty whimbrels and, on May 18, 1978, no fewer than 4,000 dunlins. Rarer species included willets, marbled godwits, avocets, ruffs, and all three species of phalarope. We also saw large flocks of other water birds, including 15,000 herring gulls in 1974 and 7,000 common terns in 1976.

Appendix A

Mammals, Reptiles & Amphibians, and Plants of the Marshes

Mammals

When northwestern Ohio was settled, the larger mammals, especially the predators, were soon exterminated. Some were killed for food or hides, others because they preyed on livestock. Harold Mayfield[1] has listed the approximate years when they disappeared:

bison	1812	gray wolf	1860
elk	1822	black bear	1860
beaver	1837	bobcat	1878
wolverine	1842	porcupine	1874
panther	1845	deer	1889
lynx	1848	otter	1900

As the forests were razed, about the middle 1850s, gray squirrels and gray foxes were replaced by fox squirrels and red foxes. In 1870, Potter reported "red, black, and silver-gray or cross foxes."[2] Red foxes were already present in early days, however: a 1782 account of fur sales in Toledo mentions 700 red fox skins.[3] These foxes have undergone interesting changes. Moving into Lucas and Ottawa counties with the clearing of the land, they apparently remained numerous until about 1900 and then declined rapidly. I do not recall that they were ever found here from 1910 until about 1935, but then they climbed swiftly to a peak in 1956. Since then, their numbers have stabilized at a point probably equal to that of the late 1800s. Gray foxes also increased in the 1940s, but have since dropped to a very limited population.

White-tailed deer had disappeared by 1889, but began to return about 1930. Today in the marshes there are probably five hundred to one thousand, thanks largely to the dikes; dikes have also benefited woodchucks, opossum, skunk, red fox, voles and shrews. The controlled water levels also favor muskrats, but they were undoubtedly already abundant by 1900. Apparently raccoons have always been numerous, a fact that was mentioned as early as 1782.

Through the years, the sale of muskrat hides has financed many of the improvements made in the marshes. In 1921, the Cedar Point Marsh owners sold 11,757 hides for $30,000. During the 1920s, Magee Marsh muskrats brought in from $5,000 to $37,000 a year; in the eleven years between the winters of 1939–40 and 1950–51, 96,900 muskrats were trapped in Magee. In 1972, 4,002 muskrats were trapped in the Cedar Point Marsh and 5,287 in the remainder of the Ottawa National Wildlife Refuge. Muskrats comprise more than seventy percent of all Ohio fur sales and nearly two-thirds of the cash value of those sales, which totals $750,000 to $1,500,000.[4] Until 1973, average fur prices showed little variation through the years compared with other costs, as the following price chart shows. In the last few years, however, raccoon and mink have more than doubled, then dropped again, perhaps because of campaigns against the wearing of furs.

Fur Prices

	1870	1966	1973	1978	1990
Muskrat	$0.16	$1.69	$2.78	$6.05-6.40	$0.75-1.23
Raccoon	$0.80	$2.82	$10.10	$45-62	$6.00-12.50
Mink	$8.00	$6.51	$20.35	$21-35	$15.00-25.00
Skunk	$0.80	$0.83	$1.97	$2.50-3.50	—
Opossum	$0.08	$0.49	$2.27	$4.75-9.00	$0.50
Weasel	—	$0.84	—	$1.00	—
Red Fox	$1.25	$6.13	$27.40	$70-89	$10.00-18.00
Gray Fox	$0.75	$2.03	$14.56	$50-62.50	$8.00-10.00
Badger	$0.60	$1.75	—	$37	—
Otter	$5.50	—	—	—	—
Bobcat	$0.50	—	—	$115 top	—
Gray Wolf	$1.50	—	—	—	—

An interesting item in the 1870 fur sales record is house cat hides at twenty cents.

Even if the pelts had no sale value, the fur-bearers now living in the marshes would have to be controlled in some way, both to maintain a habitat equally favorable to all wildlife and to minimize the damage done to dikes by muskrats, woodchucks, and other burrowing mammals.

The following list of the mammals found in Lucas and Ottawa Counties is taken from a 1981 list of mammals found throughout Ohio, compiled by Jack L. Gottschang.[5] The estimated frequency of occurrence in column 1 was given to me in 1938 by Dr. Edward S. Thomas, Curator of Natural History at the Ohio State Museum; column 2 is based on Gottschang's estimations; and column 3 represents my own observations of the mammals living in or near the marshes themselves. The term "general" means that the mammal is found throughout Ohio.

Mammals of Northwest Ohio

	Lucas/Ottawa Co.		Marshes
	1938	1981	1990
Common Opossum (*Didelphis virginiana*)	Common	Common	Common
Smoky Shrew (*Sorex fumeus*)	Uncommon	None	None
Short-tailed Shrew (*Blarina brevicauda*)	General	Common	Common
Least Shrew (*Cryptotis parva*)	General	F.Common	F.Common
Eastern Mole (*Scalopus aquaticus*)	Common	Common	None
Star-nosed Mole (*Condylura cristata*)	Rare	None	None
Little Brown Bat (*Myotis lucifugus*)	Abundant	F.Common	F.Common
Keen's Myotis (*Myotis keenii*)	Rare	Rare	None
Indiana Bat (*Myotis sodalis*)	Rare	Rare	None
Silver-haired Bat (*Lasionycteris noctivagans*)	Rare	Rare	Rare
Georgian Bat (*Pipistrellus subflavus*)	Common	Rare	None
Big Brown Bat (*Eptesicus fuscus*)	Common	Common	F.Common
Evening Bat (*Nycticeius humeralis*)	Rare	None	None

	Lucas/Ottawa Co.		Marshes
	1938	*1981*	*1990*
Red Bat			
(*Lasiurus borealis*)	General	F.Common	Rare
Hoary Bat			
(*Lasiurus cinereus*)	Rare	Rare	Rare
Raccoon			
(*Procyon lotor*)	Common	Common	Common
Striped Skunk			
(*Mephitis mephitis*)	Common	Common	Common
Least Weasel			
(*Mustela nivalis*)	Rare	Uncommon	None
Long-tailed Weasel			
(*Mustela frenata*)	General	Rare	None
Big Brown Mink			
(*Mustela vison*)	General	Few	F.Common
Red Fox			
(*Vulpes vulpes*)	General	Common	Common
Gray Fox			
(*Urocyon cinereoargenteus*)	Uncommon	Rare	Rare
Coyote			
(*Canis latrans*)	Rare	Rare	Rare
Woodchuck			
(*Marmota monax*)	Common	Common	Common
Eastern Chipmunk			
(*Tamias striatus*)	Common	Common	Few
Red Squirrel			
(*Tamiasciurus hudsonicus*)	Common	Common	Few
Fox Squirrel			
(*Sciurus niger*)	Common	Common	Rare
Gray Squirrel			
(*Sciurus carolinensis*)	Rare	None	None
Flying Squirrel			
(*Glaucomys volans*)	Uncommon	Uncommon	Rare
Striped Gopher			
(*Spermophilus*			
tridecemlineatus)	Rare	None	None
Muskrat			
(*Ondatra zibethicus*)	Common	Common	Abundant
So. Bog Lemming			
(*Synaptomys cooperi*)	Uncommon	Rare	None
Meadow Vole			
(*Microtus pennsylvanicus*)	Abundant	Abundant	Abundant

| | Lucas/Ottawa Co. | | Marshes |
	1938	1981	1990
Woodland Vole (*Microtus pinetorum*)	Rare	None	None
White-footed Mouse (*Peromyscus leucopus*)	General	Common	Common
Deer Mouse (*Peromyscus maniculatus*)	Rare	Common	Common
House Mouse (*Mus musculus*)	Common	Common	Common
Norway Rat (*Rattus norvegicus*)	Common	Common	F.Common
Jumping Mouse (*Zapus hudsonius*)	Rare	Uncommon	None
Eastern Cottontail (*Sylvilagus floridanus*)	Common	Common	Common
White-tailed Deer (*Odocoileus virginianus*)	Rare	Common	Common
Badger (*Taxidea taxus*)	Rare	Rare	None

During the years 1987 to 1990, mink increased greatly in the marshes, and can now be considered fairly common. However, for some unknown reason, muskrats decreased in late 1989 and 1990.

Mammals that once lived in the vicinity of the marshes:

Black Bear (*Ursus americanus*)
American Otter (*Lutra canadensis*)
Wolverine (*Gulo luscus*)
Fisher (*Martes pennanti*)
Timber Wolf (*Canis lupus*)
Canada Lynx (*Lynx canadensis*)
Wildcat (*Lynx rufus*)
Cougar (*Felis concolor*)
American Beaver (*Castor canadensis*)
Black Rat (introduced) (*Rattus rattus*)
Canada Porcupine (*Erethizon dorsatum*)
Showshoe Rabbit (*Lepus americanus*)
American Elk (*Cervus elaphus*)

Moose (*Alces alces*)
Bison (*Bison bison*)

Reptiles and Amphibians

The Lake Erie Marshes and their borders form one of the last refuges of reptiles and amphibians in northwestern Ohio. They were once so abundant that a mosaic picturing a frog decorates the entrance to the Lucas County Courthouse, and tales of monstrous snakes appear in early writings. Their great decline is attributed to drainage of small marshes and wet meadows; the use of insect sprays and herbicides deadly to amphibians, as well as chemical fertilizers, which interfere with amphibians' ability to breed; and uncontrolled collecting for food, pets, or profit.

Particularly distressing is the wanton destruction of snakes by people who "just don't like them." One May day several years ago, as I was concentrating on the warblers along the bird trail in Magee Marsh, I became aware of a middle-aged woman approaching, accompanied by two boys about twelve and fourteen years old. She was carrying a heavy stick as big as a cane. The trio stopped beside a fallen tree and the woman said, "Pick out a nice heavy stick like mine. We are going to kill every snake we find on the trail." I didn't say a word, and have always regretted my silence.

Lizards and Snakes

In 1951, Roger Conant recorded the following reptiles, found in the Lake Erie marshes:[6]

—**Five-lined skink** (*Eomeces fasiatus*). This is the only lizard found in northwestern Ohio. Formerly widespread, in recent years it appears to be confined to the outer beaches of the Lake Erie marshes, especially where there are piles of driftwood.

—**Blue Racer** (*Coluber constrictor flaviventris*). Specimens of this fast-moving snake have been collected from the Cedar Point Marsh and the peninsulas of Catawba and Marblehead. They were no doubt more common in the 1940s, before wooded marsh borders were cleared for farming.

—**Fox Snake** (*Elaphe vulpina*). Although not as numerous as it once was, this handsome reptile is still found regularly. In Ohio it is confined to wetlands beside Lake Erie and its tributaries. Conant's longest specimen, which came from the Cedar Point Marsh, measured fifty-nine inches, and newly hatched young are ten to eleven inches long. Because of its large size and slow reaction to humans, when this snake wanders away from the marshes it is often killed on sight. Actually, it is gentler than a water snake.

—**Milk Snake** (*Lampropeltis doliata triangulum*). This pleasingly patterned reptile was fairly common and widely distributed in both counties, but no specimens were found in or adjoining the marshes, although they probably did live there. Closest recorded sightings were from Marblehead, Catawba Island, and Genoa.

—**Kirtland's Snake** (*Natrix Kirtlandii*). Conant records this species from the wooded base of the Cedar Point Marsh and the Sandusky Bay marshes. I found one in the former area late in 1969 and another in the spring of 1975.

—**Queen Snake** (*Natrix septemvittata*). This species has been found in Erie Township and at Port Clinton, and is probably therefore an uncommon resident of the Lake Erie marshes, where it was found in 1973. It was, and perhaps still is, most numerous on the Maumee River Rapids. It has been reported recently from the Point Place area in Toledo.

—**Common Water Snake** (*Natrix sipedon sipedon*). This snake is very numerous in the marshes, where it feeds on many animals, including small fishes. Younger individuals are mottled, gradually turning black with age. Water snakes are wary and bite when cornered. The largest specimens are forty-one inches long.

—**Northern Brown Snake (DeKay's Snake)** (*Storeria dekayii dekayii*). This species is numerous on the many dikes. It is one of the gentlest snakes. Young specimens have a conspicuous yellow band across the neck resembling that of a ring-necked snake.

—**Common Garter Snake** (*Thamnophis sirtalis sirtalis*). This is one of the most common species near the marshes. Melanistic (black) individuals are common, and both types may occur in the same litter. The black type is rare throughout the United States and as a result local specimens suffer from commercial collectors.

—**Butler's Garter Snake** (*Thamnophis butleri*). This species is more common in the Lake Erie marshes and their borders than anywhere else in Ohio. It is often confused with the common garter snake.

—**Ribbon Snake** (*Thamnophis sauritis sauritis*). This slender, furtive snake prefers the fossil beaches of the Oak Openings, west of Toledo. It was not listed in the Lake Erie marshes by Conant, but was found in the Ottawa National Wildlife Refuge late in March, 1975, by Robert Crofts. Conant reports finding it in the Openings, in Sandusky, and at Cedar Point near Sandusky.

—**Hog-nosed Snake** (*Heterodon contortrix*). This alarming but harmless reptile has been collected in Ottawa County at Bay Point and Put-in-Bay. When confronted by a human, a hog-nose raises its head, opens its mouth wide, and broadens its head and neck. This threatening posture often results in its destruction. According to Karl Bednarik, hog-nose snakes were seen near Magee Marsh in 1951, but they are most common in the fossil beaches west of Toledo.[7]

Turtles

As might be expected, there are many turtles in the Lake Erie marshes, but comparatively few species.

—**Snapping Turtle** (*Chelydra serpentina*). The largest turtle, sometimes reaching forty pounds, is fairly common throughout the area. It is a very ancient form, originating about one hundred million years ago. The female deposits as many as fifty eggs in June, but these are preyed upon by raccoons, skunks, and, to a lesser extent in recent years, by crows. Newly hatched young are surprisingly small, measuring 1.1 by 1.25 inches. For several years, snappers were trapped in the Cedar Point Marsh and sold for food. In 1972, 849 were harvested.

—**Musk Turtle** (*Sternotherus odoratus*). This foul-smelling, bad-tempered, secretive species may occur in the marshes, but no specimens have been taken there. Conant lists specimens from drainage ditches west of Toledo, from Port Clinton, and from Cedar Point near Sandusky.

—**Map Turtle** (*Graptemys geographica*). This turtle prefers rivers, creeks, and the larger canals and bays in the marshes. It is wary and often confused with the painted turtle at a distance. Conant lists specimens from the Bono area and the Toussaint River. I have found them recently in the Cedar Point Marsh and the Ottawa National Wildlife Refuge. State-wide, they are most numerous in Lucas, Ottawa, and Erie counties.

—**Painted Turtle** (*Chrysemys belii marginata*). This familiar turtle is undoubtedly the most common in the marshes, in spite of the destruction of many of its nests by raccoons. Eggs are laid on dikes, even dikes surfaced with stone or gravel, and on the drier portions of the sand beaches. This species is the first to emerge in spring and the last to hibernate in fall.

—**Blanding's Turtle** (*Emys blandingii*). According to Conant, these turtles are confined almost entirely to the northwestern quarter of Ohio, especially the marshes along Lake Erie and its tributaries where they are common. A few live in the more permanent ponds and drainage ditches in the Oak Openings west of Toledo. They are docile, unsuspicious, and easily captured.

—**Spiny Soft-shelled Turtle** (*Amyda spinifera*). Only two specimens have been taken in the Lake Erie Marsh area, both of them before 1932—one at Reno Beach, the other at the mouth of Crane Creek. Since they were once numerous in Swan Creek and on the Maumee River Rapids, the absence of records may be due either to their wariness and swiftness or to a marked early decline in their numbers. More elusive than other turtles, they may still occur in the Toussaint and Portage Rivers and in the larger creeks.

—**Spotted Turtle** (*Clemmys guttata*) and **Box Turtle** (*Terapene carolina*). Conant records these turtles from Lucas County, but they are not likely to be found in the Lake Erie marshes. With one exception, a specimen found near Castalia, the Spotted Turtle has been seen only in the wet prairies of the Oak Openings. The Box Turtle is confined to the fossil sand beaches west of Toledo.

Toads and Frogs

The only complete study of Ohio toads and frogs was made by Charles F. Walker in 1946.[8] Since his collection from Lucas and Ottawa counties was mainly limited to the more unusual species, many that are common in the Lake Erie marshes are not listed. Except in the marshes and in some areas of the Oak Openings, numbers of both toads and frogs have declined greatly in recent years.

—The **American Toad** (*Bufo americanus americanus*) is found in the shallow ponds of the marshes, although it is much less common now than in the past, possibly because of the high lake levels.

—**Fowler's Toad** (*Bufo woodhousii fowleri*) is locally most numerous in the sandy areas west of Toledo; specimens have also been taken near Port Clinton and at Marblehead. Its voice is notably raucous.

—The **Cricket Frog** (*Acris crepitans*) is widespread in the marshes, but not numerous.

—The **Striped Chorus Frog** (*Pseudacris nigrita triseriata*) is an early and common vocalizer along the marsh borders, sometimes singing during January thaws.

—The **Spring Peeper** (*Hyla crucifer crucifer*) is a woodland type, greatly reduced by the clearing of the forest borders. A few still persist.

—The **Tree Toad** (*Hyla versicolor versicolor*) is a small climber, probably widespread before the swamp forests were timbered, but not likely to be found near the marshes today. It does persist in the swamp forests of the Oak Openings Metropark and similar areas.

—The **Bullfrog** (*Rana catesbeiana*), the largest of local frogs, is now numerous only in the marshes beside Lake Erie and its feeders, and in parts of the Toledo Metropark system. Stabilized ponds, built by personnel of the Ottawa National Wildlife Refuge and Magee Marsh, have been a great aid in preserving these frogs.

—The **Green Frog** (*Rana clamitans*) is a small replica of the bullfrog, widespread in the marshes but not numerous.

—The **Leopard Frog** (*Rana pipiens*) is by far the most numerous frog in the marshes. After a good hatch, they can be found every few feet on the dike banks. They have many enemies—bird, mammal, and reptile—but they endure.

—The **Northern Wood Frog** (*Rana sylvatica cantagrigensis*) is found today in the Oak Openings swamp forest, and is therefore assumed to have once lived within the marsh borders. No specimens have been taken there, however.

Salamanders

Information on salamanders is based almost entirely on my own records.

—The **Mudpuppy** (*Necturus maculosus*) is the largest area salamander. Common in the Port Clinton–Marblehead section of Lake Erie and around the islands, it is also found, in fewer numbers, in the various rivers and creeks flowing through the marshes between Toledo and Port Clinton.

—The **Jefferson Salamander** (*Ambystoma jeffersonianum*) has been found in western Lucas County and in Toledo's Detwiler Marsh, on Maumee Bay. James Imke and his son James Jr. saw several in a pond on Catawba Island south of Sugar Rock. The most recent specimen was uncovered in late January, 1974. It is a good prospect for the Lake Erie marshes.

—The **Spotted Salamander** (*Ambystoma maculatum*) has been recorded from the Swan Creek and Ottawa River valleys and from Detwiler Marsh. No specimens have been taken in the other marsh areas, but it is a good possibility.

—The **Red-spotted Newt** (*Diemictylus viridenscens viridenscens*) is a brightly colored species that has been found in the Oak Openings in both its terrestrial and aquatic stages. The Imkes have also discovered it on Catawba Island, suggesting the possibility of its occurrence in the marshes.

—**Red-backed** and **Lead-backed Salamanders** (*Plethodon cinerus cinerus*) have been discovered in both color phases in Detwiler Marsh, and the red-backed has been seen near Marblehead. George Fifer of Morrin's Point, Michigan and his son John have discovered the lead-backed phase at Crane Creek State Park in Magee Marsh within the past few years. These salamanders probably occur in some of the other Lake Erie marshes.

Marsh Plants

Wetlands vegetation is divided into four main categories, according to water level and the degree of soil saturation. These wetland types are swamp forest, shrub wetland, emergent wetland or shallow marsh, and aquatic bed or deep marsh. Lakeside marshes lie behind barrier beaches, which have their own plant associations.

An estimated 1,200 species of plants can be found in the Lake Erie marshes and their borders, including thirty shrubs and seventy-five trees. About a quarter of the plant species are not native to the region. The distribution of plant types in today's marshes is somewhat different from what it was hundreds of years ago, before Lake Erie was confined. In the unmanaged marshes, the variety of plant species was undoubtedly limited by changing water levels and the dominance of cattail and cane. Stabilizing the shoreline and constructing inland dikes multiplied the number of plant types by providing a stable shallow marsh and an additional dry-soil habitat. On the other hand, shore erosion and stone breakwaters have shrunk the sand beaches and diminished the variety of their flora.

A comprehensive list of Western Lake Erie marsh flora appears in *The Ecology of the Coastal Marshes of Western Lake Erie: A Community Profile,* by C. E. Herdendorf. This report is available in area libraries, and can be ordered free of charge from the U.S. Fish and Wildlife Service, Washington, D.C. A useful general guide to freshwater marsh plants is the illustrated *Field Guide to Nontidal Wetland Identification,* by Ralph W. Tiner, Jr.[9]

The following list of major plant species in the Lake Erie marshes and their borders was compiled by Jeanne Hawkins. I have associated them with the wetland habitats where they are most likely to be found.

Woodlands, Shrub Wetlands:

    Poison Ivy (*Rhus radicans*)
    Silky Dogwood (*Cornus obliqua*)
    Red Osier Dogwood (*Cornus stolonifera*)
    Sycamore (*Platanus occidentalis*)
    Cottonwood (*Populus deltoides*)
    Crowfoot (*Ranunculus abortivus*)
    Virginia Thimbleweed (*Anemone virginiana*)
    Columbine (*Aquilegia canadensis*)
    Wild Plum (*Prunus americana*)
    Prickly Gooseberry (*Ribes cynosbati*)

Virginia Creeper (*Parthenocissus quinquefolia*)
Large Blue Lobelia (*Lobelia siphilitica*)
Spice Bush (*Benzoin aestivale*)
Buttonbush (*Cephalanthus occidentalis*)
Virginia Dragonhead (*Physostegia virginiana*)
Carolina Ash (*Fraxinus caroliniana*)
Spotted Touch-me-not (*Impatiens biflora*)
Pale Touch-me-not (*Impatiens pallida*)
Asters (Family *Compositae, four species*)
Indian Hemp (*Apocynum cannabinum*)
Clasping-leaf Dogbane (*Apocynum medium*)
Marsh Thistle (*Cirsium palustre*)
Joe-pye Weed (*Eupatorium maculatum*)
Moonseed (*Menispermum canadensis*)
Virgin's Bower (*Clematis virginiana*)
Purple Meadow Rue (*Thalictrum dasycarpum*)
Wrinkle-leafed Goldenrod (*Solidago rugosa*)
Star Cucumber (*Sicyos angulatus*)
Swamp Milkweed (*Asclepias incarnata*)
Giant Sunflower (*Helianthus giganteus*)
Marsh Rose (*Rosa palustris*)
Pokeberry, also called pokeweed, poke, skoke, or
    pigeon-berry (*Phytolacca americana*)
Boneset (*Eupatorium perfoliatum*)
Wood-sage (*Teucrium canadense*)

Deep Marsh (aquatic plants with floating or submerged leaves):

Bladderwort (*Utricularia vulgaris*)
Water Lily (*Nymphaea tuberosa*)
Spatterdock (*Nuphar advena*)
American Lotus (*Nelumbo lutea*)
Clammy Weed (*Polinisia graveolans*)
Sweet Flag (*Acorus calamus*)
Flowering Rush (*Butomus umbellatus*)
Duckweed (*Lemna minor*)
Floating Pondweed (*Potamogeton natans*)
Water Shield (*Brasenia schreberi*)

Shallow Marsh and Wet Meadow (emergent plants with roots and
lower stems in water):

Rush (Family *Juncaceae,* 13 species)
Wild Iris (*Iris versicolor*)
Dodder (*Cuscuta gronovii*)
Broadleaf Arrowhead (*Sagittaria latifolia*)
Big Bluestem (*Andropogen gerardi*)
Bluejoint Grass (*Calamagrostis canadencis*)
Common Reed or Cane (*Phragmites*)
Pickerel Weed (*Pontederia cordata*)
Sword Grass (*Seirpus americanus*)
Coneflower (*Centaurea cyanus*)
Sweet Flag (*Acorus calamus*)
Cattail, Narrow-leaved (*Typha angusifolia*)
Cattail, Broad-leaved (*Typha latifolia*)
Monkey Flower (*Mimulus ringens*)
Blue Vervain (*Verbena hastata*)
Wrinkle-leafed Goldenrod (*Solidago rugosa*)
Cardinal Lobelia (*Lobelia cardinalis*)
Large Blue Lobelia (*Lobelia siphilitica*)
Bur Marigold (*Bidens laevis*)
Rose Mallow (*Hibiscus moscheutos*)
Sneezeweed (*Helenium nudiflorum*)
Purple Loosestrife (*Lythrum salicaria*)
Silverweed (*Potentilla anserina*)
Boltonia (*Boltonia asteroides*)
Water hemlock (*Circuta maculata*)

Dikes:

Common Greenbriar (*Smilax rotundifolia*)
Staghorn Sumac (*Rhus typhina*)
Common Milkweed (*Asclepias syriaca*)
Elderberry (*Sambucus canadensis*)
Wild Wormwood (*Artemisia biennis*)
Asters
Wild Coreopsis (*Coreopsis tripteris*)
Swamp Thistle (*Cirsium palustre*)
Coneflower (*Centaures cyanus*)
Canada Goldenrod (*Solidago canadensis*)
Yellow Goats-beard *Tragopogon pratensis*
Salsify, or Oyster Plant (*Tragopogon porrifolius*)
Bindweed, Hedge (*Convolvulus sepium*)

Bindweed, Upright (*Convolvulus spithamaeus*)
Tall Ironweed (*Vernonia altissima*)
Wild Geranium (*Geranium maculatum*)
Rough Cinquefoil (*Potentilla norvegica*)
Cottonwood (*Populus deltoides*)
Willow, Sandbar (*Salix interior*)
Stinging Nettle (*Urtica dioica*)
River Grape (*Vitis riparis*)
Catnip (*Nepeta cataria*)
Giant Ragweed (*Ambrosia trifida*)
Yarrow (*Achillea millefolium*)
Common Smartweed (*Polygonum hydropiper*)
Velvet leaf (*Abutilon theophrasti*)
Common burdock (*Arctium minus*)
Cocklebur (*Xanthium chinense*)
Wild carrot (Queen Anne's lace) (*Daucus carota*)
Evening primrose (*Oenothera binnes*)

Appendix B

Birds of the
Ohio Lake Erie Marshes

From 1926 through 1988, 361 species have been reported in the Toledo area; 167 of these nested in or near the marshes, and 181 have been seen in winter. Sixty are listed as very rare. Birds that do not appear in all seasons are migrants. The terms used have the following meanings: common—seen on 90% of visits; fairly common—seen on more than 50% of visits; uncommon—seen on 20-50% of visits; rare—seen on fewer than 20% of visits.

	Spring	*Summer*	*Fall*	*Winter*
Red-throated Loon (*Gavia stellata*)	Rare		Rare	Rare
Common Loon (*Gavia immer*)	Few		Few	Rare
Pied-billed Grebe (*Podilymbus podiceps*)	Common	Common	Common	Few
Horned Grebe (*Podiceps auritus*)	Common		Common	Few
Red-necked Grebe (*Podiceps grisegena*)	V.Rare			V.Rare
Eared Grebe (*Podiceps nigricollis*)	Rare			V.Rare
American White Pelican (*Pelicanus erythrorhynchos*)	Rare	Few	Few	

	Spring	Summer	Fall	Winter
Double-crested Cormorant (*Phalacrocorax auritus*)	Uncommon	Few	Common	Rare
American Bittern (*Botaurus lentiginosus*)	Uncommon	Uncommon	Uncommon	Rare
Least Bittern (*Ixobrychus exilis*)	Uncommon	Uncommon	Uncommon	
Great Blue Heron (*Ardea herodias*)	Abundant	Common	Common	Few
Great Egret (*Casmerodius albus*)	Common	Common	Common	Rare
Snowy Egret (*Egretta thula*)	Few	Few	Few	
Little Blue Heron (*Florida caerulea*)	Few	Few	Few	
Tricolored Heron (*Hydranassa tricolor*)	Uncommon	Uncommon		
Cattle Egret (*Bubuleus ibis*)	Common	Few	Common	V.Rare
Green-backed Heron (*Butorides striatus*)	Common	Common	Common	
Black-crowned Night Heron (*Nycticorax nycticorax*)	Common	Common	Common	Few
Yellow-crowned Night Heron (*Nyctanassa violacea*)	Uncommon	Rare	Uncommon	
Glossy Ibis (*Plegadis falcinellus*)	Rare	Rare	Rare	
Tundra Swan (*Olor columbianus*)	Common		Common	Few
Mute Swan (*Cygnus olor*)	Few	Few	Few	Few
Greater White-fronted Goose (*Anser albifrons*)	Rare		Rare	Rare

	Spring	Summer	Fall	Winter
Snow Goose (White) (*Chen caerulescens*)	Uncommon		Uncommon	Few
Snow Goose (Blue color phase)	Common		Common	Few
Brant (*Branta bernicla*)	Rare		Rare	
Canada Goose (*Branta canadensis*)	Abundant	Common	Abundant	Abundant
Canada Goose (Hutchin's) (*B. canadensis hutchinsii*)	V.Rare		Rare	Rare
Wood Duck (*Aix sponsa*)	Common	Common	Abundant	Few
Green-winged Teal (*Anas crecca*)	Abundant	Uncommon	Abundant	Few
American Black Duck (*Anas rubripes*)	Abundant	Uncommon	Abundant	Abundant
Mallard (*Anas platryhynchos*)	Abundant	Abundant	Abundant	Abundant
Northern Pintail (*Anas acuta*)	Abundant	Few	Abundant	Few
Blue-winged Teal (*Anas discors*)	Common	Common	Common	Rare
Northern Shoveler (*Anas clypeata*)	Common	Common	Common	Few
Gadwall (*Anas strepera*)	Uncommon	Few	Uncommon	Few
Eurasian Wigeon (*Anas penelope*)	Few		Rare	V.Rare
American Wigeon (*Anas americana*)	Abundant	Few	Abundant	Few
Canvasback (*Aythya valisineria*)	Abundant	Rare	Common	Abundant
Redhead (*Aythya americana*)	Common	Few	Common	Common
Ring-necked Duck (*Aythya collaris*)	Common	Few	F.Common	Rare
Greater Scaup (*Aythya marila*)	Few		Few	Few

	Spring	Summer	Fall	Winter
Lesser Scaup				
(*Aythya affinis*)	Abundant	Few	Common	Abundant
Old Squaw				
(*Clangula hyemalis*)	Uncommon		Uncommon	Uncommon
Black Scoter				
(*Melanitta nigra*)	Rare		Rare	Rare
Surf Scoter				
(*Melanitta perspicillata*)	Rare		Uncommon	Uncommon
White-winged Scoter				
(*Melanitta deglandi*)	Few		Few	Few
Common Goldeneye				
(*Bucephala clangula*)	Abundant		Common	Abundant
Bufflehead				
(*Bucephala albeola*)	Common		Common	Common
Hooded Merganser				
(*Lophodytes cucuilatus*)	Common	Few	Common	Uncommon
Common Merganser				
(*Mergus merganser*)	Abundant		F.Common	Abundant
Red-breasted Merganser				
(*M. serrator*)	Common		Common	F.Common
Ruddy Duck				
(*Oxyura jamaicensis*)	Common	Rare	Common	F.Common
Turkey Vulture				
(*Cathartes aura*)	Abundant	F.Common	Common	Rare
Osprey				
(*Pandion haliaetus*)	F.Common	Rare	F.Common	V.Rare
Northern Harrier				
(*Circus cyaneus*)	Common	Rare	Common	F.Common
Sharp-shinned Hawk				
(*Accipiter striatus*)	Common	Rare	Common	Uncommon
Cooper's Hawk				
(*Accipiter cooperii*)	Common	Uncommon	Common	Uncommon
Northern Goshawk				
(*Accipiter gentilis*)	Rare		Rare	Rare
Red-shouldered Hawk				
(*Buteo lineatus*)	Common	Uncommon	Common	Few
Broad-winged Hawk				
(*Buteo platypterus*)	Common		Common	

Appendix B

	Spring	Summer	Fall	Winter
Red-tailed Hawk (*Buteo jamaicensis*)	Abundant	Common	Common	Common
Rough-legged Hawk (*Buteo lagopus*)	F.Common		Uncommon	Common
Golden Eagle (*Aquila chrysaetus*)	Rare		Rare	Rare
American Kestrel (*Falco sparverius*)	Common	Common	Common	Common
Merlin (*Falco columbarius*)	Rare		Rare	Rare
Peregrine Falcon (*Falco peregrinus*)	Few	Rare	Few	Few
Gyrfalcon (*Falco rusticolus*)	V.Rare		V.Rare	V.Rare
Ring-necked Pheasant (*Phasianus colchicus*)	Common	Common	Common	Common
Northern Bobwhite (*Colinus virginianus*)	Rare	Rare	Rare	Rare
Yellow Rail (*Coturnicops noveboracensis*)	V.Rare		V.Rare	
King Rail (*Rallus elegans*)	Uncommon	Uncommon	Uncommon	V.Rare
Virginia Rail (*Rallus limicola*)	Few	Few	Few	Rare
Sora (*Porzana carolina*)	Few	Uncommon	Common	Rare
Common Moorhen (*Gallinula chloropus*)	F.Common	F.Common	F.Common	V.Rare
American Coot (*Fulica americana*)	Abundant	Common	Abundant	Few
Sandhill Crane (*Grus canadensis*)	Rare		Rare	Rare
Black-bellied Plover (*Pluvialis squatarola*)	Abundant	V.Rare	Common	
Lesser Golden Plover (*Pluvialis dominica*)	Abundant		Common	

	Spring	Summer	Fall	Winter
Semi-palmated Plover (*Charadrius semipalmatus*)	Common	Rare	Common	
Piping Plover (*Charadrius melodus*)	V.Rare		V.Rare	
Killdeer (*Charadrius vociferus*)	Abundant	Common	Abundant	Uncommon
American Avocet (*Recurviorstra americana*)	Uncommon		Uncommon	
Greater Yellowlegs (*Tringa melanoleuca*)	Common	Rare	Common	V.Rare
Lesser Yellowlegs (*Tringa flavipes*)	Abundant	Rare	Abundant	V.Rare
Solitary Sandpiper (*Tringa soliteria*)	Common		Common	
Willet (*Catoptrophorus semipalmatus*)	Uncommon		Uncommon	
Spotted Sandpiper (*Actitis maculbria*)	Common	Common	Common	
Upland Sandpiper (*Bartramia longicauda*)	Few	Few	Few	
Whimbrel (*Numenius phaeopus*)	Few		Few	
Hudsonian Godwit (*Limosa haemastica*)	Few		Few	
Marbled Godwit (*Limosa fedoa*)	Few		Few	
Ruddy Turnstone (*Arenaria interpres*)	Common	Rare	Common	Rare
Red Knot (*Calidris canutus*)	Few	Rare	Few	
Sanderling (*Calidris alba*)	Common	Rare	Common	Uncommon

	Spring	Summer	Fall	Winter
Semipalmated Sandpiper (*Calidris pusilla*)	Common	Rare	Abundant	
Western Sandpiper (*Calidris mauri*)	Few		Few	
Least Sandpiper (*Calidris minutilla*)	Common		Abundant	
White-rumped Sandpiper (*C. fuscicollis*)	Few		Few	
Baird's Sandpiper (*Calidris bairdii*)	Few		Few	Rare
Pectoral Sandpiper (*Calidris melanotus*)	Abundant	Rare	Abundant	
Purple Sandpiper (*Calidris maritima*)	V.Rare		V.Rare	V.Rare
Dunlin (*Calidris alpina*)	Abundant	Uncommon	Abundant	Uncommon
Stilt Sandpiper (*Micropalama himantopus*)	Uncommon		Common	
Buff-breasted Sandpiper (*Tryngites subruficorlis*)	Rare		Uncommon	
Ruff (*Philomachus pugnax*)	Rare		Rare	
Short-billed Dowitcher (*Limnodromus griseus*)	Common		Common	
Long-billed Dowitcher (*L. scolopaceus*)	Uncommon		Common	
Common Snipe (*Capella gallinago*)	Common	Few	Common	Uncommon
American Woodcock (*Philohela minor*)	F.Common	F.Common	Common	Rare
Wilson's Phalarope (*Steganopus tricolor*)	Common	V.Rare	Common	

	Spring	Summer	Fall	Winter
Red-necked Phalarope (*Lobipes lobatus*)	Few		Uncommon	V.Rare
Red Phalarope (*Phalaropus fulicarius*)	Rare		Uncommon	V.Rare
Laughing Gull (*Larus africilla*)	Rare	V.Rare	Rare	
Franklin's Gull (*Larus pipixcan*)	Few	Rare	Uncommon	Rare
Little Gull (*Larus minutus*)	Rare		Rare	V.Rare
Common Black-headed Gull (*Larus ridibundus*)	V.Rare		Rare	Rare
Bonaparte's Gull (*Larus philadelphia*)	Common	Few	Abundant	Common
Ring-billed Gull (*Larus delawarensis*)	Abundant	Abundant	Abundant	Common
Herring Gull (*L. argentatus*)	Abundant	Common	Abundant	Abundant
Thayer's Gull (*L. thayeri*)	Rare		Rare	Rare
Iceland Gull (*L. glaucoides*)	Uncommon		Uncommon	Uncommon
Lesser Black-backed Gull (*Larus fuscus*)	Rare		Rare	Rare
Glaucous Gull (*Larus hyperboreus*)	Uncommon	V.Rare	Uncommon	Uncommon
Great Black-backed Gull (*Larus marinus*)	Common	Few	Common	Common
Black-legged Kittiwake (*Rissa tridactyla*)	Rare		Rare	Rare
Sabine's Gull (*Xema sabini*)	V.Rare		V.Rare	V.Rare
Caspian Tern (*Sterna caspia*)	Common	Few	Common	Few
Common Tern (*Sterna hirundo*)	Common	Few	Common	V.Rare

	Spring	Summer	Fall	Winter
Forster's Tern (*Sterna forsteri*)	Common		Common	
Least Tern (*Sterna albifrons*)	Rare		Rare	
Black Tern (*Chlidonias niger*)	Few	Few	Few	
Rock Dove (*Columba livia*)	Common	Common	Common	Common
Mourning Dove (*Zenaida macroura*)	Common	Common	Abundant	Common
Black-billed Cuckoo (*Coccyzus erythropthalmus*)	Uncommon	Uncommon	Uncommon	
Yellow-billed Cuckoo (*C. americanus*)	Common	Common	Common	
Common Barn Owl (*Tyto alba*)	Rare	Rare	Rare	Rare
Eastern Screech-Owl (*Otus asio*)	Common	Common	Common	Common
Great Horned Owl (*Bubo virginianus*)	Common	Common	Common	Common
Snowy Owl (*Nyotea scandiaca*)	Few		Few	Few
Long-eared Owl (*Asio otus*)	Uncommon	V.Rare	Uncommon	Uncommon
Short-eared Owl (*Asio flammeus*)	Uncommon	V.Rare	Uncommon	Uncommon
Northern Saw-whet Owl (*Aegolius acadicus*)	Few	Rare	Few	Rare
Common Nighthawk (*Chordeiles minor*)	Common	Uncommon	Common	
Whip-poor-will (*Caprimulgus vociferus*)	Uncommon		Rare	
Chimney Swift (*Chaetura pelagica*)	Common	Few	Common	
Ruby-throated Hummingbird (*Archilochus colubris*)	Few	Rare	Common	

	Spring	Summer	Fall	Winter
Belted Kingfisher (*Megaceryle alcyon*)	Common	Few	Common	Uncommon
Red-headed Woodpecker (*Melanerpes erythrocephalus*)	F.Common	Uncommon	Uncommon	Few
Red-bellied Woodpecker (*M. carolinus*)	Uncommon	Uncommon	Uncommon	Uncommon
Yellow-bellied Sapsucker (*Sphyrapicus varius*)	Common		Common	Uncommon
Downy Woodpecker (*Picoides pubescens*)	Common	Common	Common	Common
Hairy Woodpecker (*Picoides villosus*)	Uncommon	Uncommon	Uncommon	Uncommon
Northern Flicker (*Colaptes auratus*)	Common	Common	Common	Few
Olive-sided Flycatcher (*Nuttallornis borealis*)	Uncommon		Uncommon	
Eastern Wood Pewee (*Cuntopus virens*)	Common	Common	Common	
Yellow-bellied Flycatcher (*Empidonax flaviventris*)	Common		Common	
Acadian Flycatcher (*E. virescens*)	Few		Few	
Alder Flycatcher (*E. alnorum*)	Rare		Rare	
Willow Flycatcher (*E. traillii*)	Common	Uncommon	Common	
Least Flycatcher (*E. minimus*)	Common		Common	
Eastern Phoebe (*Sayornis phoebe*)	Common	Uncommon	Common	V.Rare
Great Crested Flycatcher (*Myarchus crinitus*)	Common	Common	Common	
Western Kingbird (*Tyrannus verticalis*)		V.Rare	Rare	

	Spring	Summer	Fall	Winter
Eastern Kingbird (*Tyrannus tyrannus*)	Common	Common	Common	
Horned Lark (*Eremophila alpestris*)	Common	Few	Common	Common
Purple Martin (*Progne subis*)	F.Common	F.Common	F.Common	
Tree Swallow (*Iridoprocne bicolor*)	Common	Common	Abundant	Rare
Northern Rough-winged Swallow (*Stelgidopteryx ruficollis*)	Uncommon	Rare	Uncommon	
Bank Swallow (*Riparia riparia*)	Common	F.Common	Abundant	
Cliff Swallow (*Petrochelidon pyrrhonota*)	Uncommon	Few	Uncommon	
Barn Swallow (*Hirundo rustica*)	Common	Common	Common	V.Rare
Blue Jay (*Cyanocitta cristata*)	Abundant	Common	Common	F.Common
American Crow (*Corvus brachyrhynchos*)	Common	Few		
Black-capped Chickadee (*Parus atricapillus*)	Uncommon		Uncommon	Uncommon
Tufted Titmouse (*Parus bicolor*)	Few	Few	Few	Few
White-breasted Nuthatch (*Sitta Carolinensis*)	Common	Uncommon	Common	Common
Red-breasted Nuthatch (*Sitta canadensis*)	Uncommon	Rare	Uncommon	Uncommon
Brown Creeper (*Certhia familiaris*)	Common		Common	Uncommon
Carolina Wren (*Thryothorus ludovicianus*)	Uncommon	Uncommon	Uncommon	Uncommon
House Wren (*Troglodytes aedon*)	Abundant	Common	Common	

	Spring	Summer	Fall	Winter
Winter Wren (*T. troglodytes*)	Common		Common	F.Common
Sedge Wren (*Cistothorus platensis*)	Rare	Rare	Rare	V.Rare
Marsh Wren (*Cistothorus palustris*)	Uncommon	Uncommon	Uncommon	Rare
Golden-crowned Kinglet (*Regulus satrapa*)	Abundant		Common	F.Common
Ruby-crowned Kinglet (*R. calendula*)	Common		Common	Few
Blue-gray Gnatcatcher (*Polioptila caerulea*)	Common	Rare	F.Common	V.Rare
Eastern Bluebird (*Sialia sialis*)	Uncommon		Rare	
Veery (*Catharus fuscescens*)	Common		Common	
Gray-cheeked Thrush (*Catharus minimus*)	Common		Common	
Swainson's Thrush (*Catharus ustulatus*)	Abundant		Abundant	
Hermit Thrush (*Catharus guttatus*)	V.Common		Common	Few
Wood Thrush (*Hylocichla mustelina*)	Common	Few	Common	V.Rare
American Robin (*Turdus migratorius*)	Abundant	Common	Abundant	Few
Gray Catbird (*Dumetella carolinensis*)	Common	Common	Common	Rare
Northern Mockingbird (*Mimus polyglottos*)	Few	Uncommon	Few	Rare

	Spring	Summer	Fall	Winter
Brown Thrasher *(Toxostoma rufum)*	Common	Common	F.Common	Rare
Water Pipit *(Anthus spinoletta)*	F.Common		F.Common	Rare
Cedar Waxwing *(Bombycilla cedrorum)*	Common	Uncommon	Common	Uncommon
Northern Shrike *(Lanius excubitor)*	Rare		Rare	Rare
Loggerhead Shrike *(Lanius ludovicianus)*	Rare	V.Rare	Rare	V.Rare
European Starling *(Sturnus vulgaris)*	Abundant	Common	Abundant	Abundant
White-eyed Vireo *(Vireo griseus)*	Few	Few	Few	
Solitary Vireo *(Vireo solitarius)*	Common		Common	
Yellow-throated Vireo *(V. flavifrons)*	Uncommon	V.Rare	Uncommon	
Warbling Vireo *(Vireo gilvus)*	Common	Common	Common	
Philadelphia Vireo *(V. philadelphicus)*	Common		Uncommon	
Red-eyed Vireo *(Vireo olivaceus)*	Common	Common	Common	
Blue-winged Warbler *(Vermivora pinus)*	Uncommon	V.Rare	Rare	
Golden-winged Warbler *(V. chrysoptera)*	Uncommon	V.Rare		
Brewster's Hybrid Warbler *(Vermivora leucobronchialis)*	Uncommon			
Lawrence's Hybrid Warbler *(Vermivora lawrencei)*	Rare			

	Spring	Summer	Fall	Winter
Tennessee Warbler (*V. peregrina*)	Common		Common	
Orange-crowned Warbler (*V. celata*)	Uncommon		Uncommon	Rare
Nashville Warbler (*V. ruficapilla*)	Common		Common	
Northern Parula Warbler (*Parula americana*)	Uncommon		Uncommon	
Yellow Warbler (*Dendroica petechia*)	Abundant	Common	F.Common	
Chestnut-sided Warbler (*Dendroica pennsylvanica*)	Common		Uncommon	
Magnolia Warbler (*Dendroica magnolia*)	Abundant		Common	
Cape May Warbler (*Dendroica tigrina*)	Abundant		Common	
Black-throated Warbler (*Dendroica caerulescens*)	Common		Common	
Yellow-rumped (Myrtle) Warbler (*D. coronata*)	Abundant		Abundant	F.Common
Black-throated Green Warbler (*D. virens*)	Common		Common	
Blackburnian Warbler (*D. fusca*)	Common		Common	
Yellow-throated Warbler (*D. dominica*)	Rare			
Pine Warbler (*D. pinus*)	Uncommon		Rare	Rare
Kirtland's Warbler (*D. kirtlandii*)	Rare		V.Rare	
Prairie Warbler (*D. discolor*)	Rare		V.Rare	

	Spring	Summer	Fall	Winter
Palm Warbler (*D. palmarum*)	Common		Common	
Yellow Palm Warbler (*D. palmarum hypochrysea*)	Uncommon		Rare	
Bay-breasted Warbler (*D. castanea*)	Common		Abundant	
Blackpoll Warbler (*D. striata*)	Common		Abundant	
Cerulean Warbler (*D. cerulea*)	Uncommon		Rare	
Black-and-White Warbler (*Mniotilta varia*)	Common		Common	
American Redstart (*Setophaga ruticilla*)	Abundant	V.Rare	Common	
Prothonotary Warbler (*Protonotaria citrea*)	Few	Few	Few	
Worm-eating Warbler (*Helmitheros vermivorus*)	Rare		Rare	
Ovenbird (*Seiurus aurocapillus*)	Common	Rare	Common	
Northern Waterthrush (*Seiurus noveboracensis*)	Common		Common	V.Rare
Louisiana Waterthrush (*S. motacilla*)	Uncommon			
Kentucky Warbler (*Oporornis formosus*)	Rare		V.Rare	
Connecticut Warbler (*Oporornis agilis*)	Uncommon		Uncommon	
Mourning Warbler (*O. philadelphia*)	Uncommon	V.Rare	Uncommon	
Common Yellowthroat (*Geothlypis trichas*)	Abundant	Common	Common	Uncommon

	Spring	Summer	Fall	Winter
Hooded Warbler (*Wilsonia citrina*)	Uncommon		Rare	
Wilson's Warbler (*Wilsonia pusilla*)	Common		Common	
Canada Warbler (*Wilsonia* *canadensis*)	Common		F.Common	
Yellow-breasted Chat (*Icteria virens*)	F.Common	Rare	Rare	
Summer Tanager (*Piranga rubra*)	Rare			
Scarlet Tanager (*Piranga olivacea*)	Common	Rare	F.Common	
Northern Cardinal (*Cardinalis* *cardinalis*)	Common	Common	Common	Common
Rose-breasted Grosbeak (*Pheucticus* *ludovicianus*)	Common		Common	
Blue Grosbeak (*Guiraca caerulea*)	Rare			
Indigo Bunting (*Passerina cyanea*)	Abundant	Common	Common	
Dickcissel (*Spiza americana*)	Few	Few	Few	
Rufous-sided Towhee (*Pipilo* *erythophthalmus*)	Common	Common	Common	Few
American Tree Sparrow (*Spizella arborea*)	Abundant		Common	Abundant
Chipping Sparrow (*Spizella passerina*)	Uncommon	Uncommon	Uncommon	V.Rare
Field Sparrow (*Spizella pusilla*)	Common	Uncommon	Common	Rare
Vesper Sparrow (*Pooecetes* *gramineus*)	Common	F.Common	Common	

	Spring	Summer	Fall	Winter
Savannah Sparrow *(Passerculus sandwichensis)*	Common	F.Common	Common	V .Rare
Grasshopper Sparrow *(Ammodramus savannarum)*	Rare	Rare	V.Rare	
Henslow's Sparrow *(A. henslowii)*	Rare	Rare	V.Rare	
LeConte's Sparrow *(A. leconteii)*	V.Rare		V.Rare	
Sharp-tailed Sparrow *(A. caudacuta)*	V.Rare		V.Rare	
Fox Sparrow *(Passerella iliaca)*	Common		Common	Uncommon
Song Sparrow *(Melospiza melodia)*	Common	Common	Common	Common
Lincoln's Sparrow *(Melospiza lincolnii)*	F.Common		F.Common	Rare
Swamp Sparrow *(M. georgiana)*	Common	Rare	Common	Common
White-throated Sparrow *(Zonotrichia albicollis)*	Abundant	V.Rare	Abundant	F.Common
White-crowned Sparrow *(Z. leucophrys)*	Common		Common	Uncommon
Dark-eyed (slate-colored) Junco *(Junco hyemalis)*	Abundant		Abundant	Common
Oregon Junco *(Junco oreganus)*	Uncommon		Uncommon	Uncommon
Lapland Longspur *(Calcarius lapponicus)*	F.Common		Uncommon	F.Common
Snow Bunting *(Plectrophenax nivalis)*	Common		Common	F.Common

	Spring	Summer	Fall	Winter
Bobolink (*Dolichonyx orzivorus*)	Uncommon	Uncommon	Uncommon	
Red-winged Blackbird (*Agelaius phoeniceus*)	Abundant	Abundant	Abundant	Common
Eastern Meadowlark (*Sturnella magna*)	Uncommon	Uncommon	Uncommon	Few
Western Meadowlark (*S. neglecta*)	Uncommon	Uncommon	Uncommon	
Yellow-headed Blackbird (*Xantho-cephalus xanthocephalus*)	Uncommon	Rare	Uncommon	Rare
Rusty Blackbird (*Euphagus carolinus*)	Common		Common	Uncommon
Brewer's Blackbird (*E. cyanocephalus*)	Rare		Rare	V.Rare
Common Grackle (*Quiscalus quiscula*)	Abundant	Abundant	Abundant	F.Common
Brown-headed Cowbird (*Molothrus ater*)	Common	Common	Common	F.Common
Orchard Oriole (*Icterus spurius*)	Uncommon	Uncommon	Few	
Northern Oriole (*Icterus galbula*)	Common	F.Common	Common	Rare
Pine Grosbeak (*Pinicola enucleator*)	Rare		V.Rare	Rare
Purple Finch (*Carpodacus purpureus*)	F.Common	Uncommon	Abundant	Common
House Finch (*Carpodacus mexicanus*)	Common	Common	Common	Common
Red Crossbill (*Loxia curvirostra*)	Rare	V.Rare	Rare	Rare
Common Redpoll (*Carduelis flammea*)	F.Common		F.Common	F.Common

	Spring	Summer	Fall	Winter
Hoary Redpoll *(Carcuelis hornemanni)*	V.Rare			V.Rare
Pine Siskin *(Carduelis pinus)*	Common		Uncommon	Few
American Goldfinch *(Carduelis tristis)*	Abundant	Common	Common	Common
Evening Grosbeak *(Hesperiphona vespertina)*	Uncommon		Uncommon	Uncommon
House Sparrow *(Passer domesticus)*	Abundant	Abundant	Abundant	Abundant

The following birds were seldom reported between 1926 and 1991:

Western Grebe *(Aechmophorus occidentalis)*
Northern Gannet *(Morus bassanus)*
Brown Pelican *(Pelecanus occidentalis)*
White-faced Ibis *(Plegadis chihi)*
Wood Stork *(Mycteria americana)*
Greater Flamingo *(Phoenicopterus ruber)*
Fulvous Whistling Duck *(Dendrocygna bicolor)*
Ross' Goose *(Chen rossii)*
Cinnamon Teal *(Anas cyanoptera)*
Common Eider *(Somateria millissima)*
King Eider *(Somateria spectabilis)*
Harlequin Duck *(Histrionicus histrionicus)*
Barrow's Goldeneye *(Bucephala islandica)*
Ruddy Shelduck *(Tadorna ferruginea)*
Mississippi Kite *(Ictinia mississippiensis)*
Swainson's Hawk *(Buteo swainsoni)*
Black Rail *(Laterallus jamaicensis)*
Purple Gallinule *(Porphyrula martinica)*
Wilson's Plover *(Charadrius wilsonia)*
Black-necked Stilt *(Himantopus mexicanus)*
Long-billed Curlew *(Numenius americanus)*
Curlew Sandpiper *(Erolia ferruginea)*
Sharp-tailed Sandpiper *(Pisobia acuminata)*
Pomarine Jaeger *(Stercorarius pomarinus)*
Parasitic Jaeger *(Stercorius parasiticus)*

Long-tailed Jaeger (*Stercorius longicaudus*)
Kumlien's Gull (*Larus glaucoides*)
Skua (*Catharacta skua*)
Ancient Murrelet (*Synthliboramphus antiquum*)
Atlantic Puffin (*Fratercula arctica*)
Ringed Turtle Dove (*Streptopelia risoria*)
Groove-billed Ani (*Crotophago ani*)
Burrowing Owl (*Athene cunicularia*)
Great Gray Owl (*Strix nubulosa*)
Black-backed Woodpecker (*Picoides arcticus*)
Red-shafted Flicker (*Colaptes auratus cafer*)
Pileated Woodpecker (*Dryocopus pileatus*)
Black-billed Magpie (*Pica pica*)
Boreal Chickadee (*Parus hudsonicus*)
Bewick's Wren (*Thryomanes bewickii*)
Northern Wheatear (*Oenanthe oenanthe*)
Townsend's Solitaire (*Myadestes townsendi*)
Bohemian Waxwing (*Bombycilla garrulus*)
Townsend's Warbler (*Dendroica townsendi*)
Macgillivray's Warbler (*Oporornis tolmiei*)
Western Tanager (*Piranga ludoviciana*)
Black-headed Grosbeak (*Pheucticus melanodephalus*)
Bachman's Sparrow (*Aimophila aestivalis*)
Clay-colored Sparrow (*Spizella pallida*)
Lark Sparrow (*Chondestes grammacus*)
Black-chinned Sparrow (*Spizella atrogularis*)
Baird's Sparrow (*Ammodramus henslowii*)
White-winged Junco (*Junco hymali aiken*)
Smith's Longspur (*Calcarius pictus*)
Great-tailed Grackle (*Quiscalus mexicanus*)
White-winged Crossbill (*Loxia leucoptera*)
Greater Redpoll (*Acanthis flammea rostrata*)
European Goldfinch (*Carduelis carduelis*)
Rosy Finch (*Leucosticte tephrocotis*)

The following waterfowl are probably escapees from captivity:

Muscovy Duck (*Cairina moschata*)
Graylag Goose (*Anser anser*) (from England)
Lesser White-fronted Goose (*Anser erythropus*)

Swan-goose (*Anser cygnoides*)
Red-breasted Goose (from Hungary)
Bar-headed Goose (*Anser indicus*)
Barnacle Goose (*Branta leucopsis*)
Pink-footed Goose (*Anser brachyrhynchus*)
Egyptian Goose (*Alopochen aegyptiacus*)
Trumpeter Swan (*Olor buccinator*)

Notes

Chapter 1

1. Richard P. Goldthwait, "Ice Over Ohio," in *Ohio's Natural Heritage,* edited by Michael B. Lafferty (Columbus: Ohio Academy of Science, 1979, 32–47), 34.

2. Ibid, 36.

3. Jane L. Forsyth, Professor of Geology, Bowling Green State University, personal communication.

4. Jane L. Forsyth, "Late-glacial and postglacial history of western Lake Erie," *The Compass of Sigma Gamma Epsilon* 51, no. 1 (1973): 16–20.

5. E. Jaworski, C. N. Raphael, P. J. Mansfield, and B. B. Williamson, "Impact of Great Lakes water level fluctuations on coastal wetlands," in *Fish and Wildlife Resources of the Great Lakes Coastal Wetlands, Vol. 1: Overview,* edited by C. E. Herdendorf, S. M. Hartley, and M. D. Barnes (Washington, D.C.: U.S. Fish and Wildlife Service, 1981), 48.

6. C. F. M. Lewis, "Late quaternary history of lake levels in the Huron and Erie Basins," *Proceedings of the 12th Conference on Great Lakes Research* (Ann Arbor, Michigan: International Association for Great Lakes Research, 1969, 250–70), 267.

7. C. E. Herdendorf and R. L. Stuckey, "Lake Erie and the Islands," in Lafferty (214–35), 222.

8. Jack L. Hough, *Geology of the Great Lakes* (Urbana: University of Illinois Press, 1958), 44.

9. "Great Lakes Basin Commission Report," *The Communicator* (April, 1973).

10. U.S. Department of Commerce, NOAA, National Ocean Survey for 1972–73. *Monthly Bulletin of Lake Levels* (Detroit: Lake Survey Center, 1973).

11. J. D. Riddell, *Report No. 60 to the Governor and Legislature of Ohio* (Columbus, Ohio: James B. Gardiner, 1837), 2.

12. Samuel Brown, *Views of the Campaigns of the Northwestern Army, &c.* (Philadelphia: William G. Murphey, 1815), 137–41.

13. Harlan Hatcher, *Lake Erie* (Indianapolis: Bobbs-Merrill, 1945).

14. C. S. Van Tassel, *History of the Maumee Valley, Toledo, and Sandusky Region* (Chicago: S. J. Clarke, 1929).

15. Ibid.

16. Ibid.

17. Captain Frank E. Hamilton, "A Forgotten Port and Log Towing Revenue Cutters," *Telescope* (Great Lakes Maritime Institute, Detroit, Michigan) 18, no.1 (1969): 207–11.

18. Josephine Fassett, *History of Oregon and Jerusalem Townships, Lucas County, Ohio, 1837–1961* (Arkansas: Camden Co., 1961), 95.

19. Brown, 140.

20. Milton B. Trautman, *The Fishes of Ohio* (Columbus: Ohio State University Press, 1957), 26–29.

21. Harry Byers, James's son, personal communication.

22. John B. Uhl, *Official Atlas of Lucas County, Ohio* (Toledo, Ohio: Uhl Bros., 1900); *Historical Atlas of Ottawa County, Ohio* (Chicago, Illinois: H. H. Hardesty, 1874); *Atlas of Ottawa County, Ohio* (H. J. Goodman, 1900).

23. Fassett, 94.

24. Ralph Andrews, "A Study of Waterfowl Nesting on a Lake Erie Marsh," (Masters Thesis, The Ohio State University, Columbus, Ohio, 1952).

Chapter 2

1. Reuben G. Thwaites, *The Jesuit Relations and Allied Documents, etc.*, vol. 69 (Cleveland, Ohio: Burrows Brothers Co., 1901), 32.

2. C. S. Van Tassel, *History of the Maumee Valley, Toledo, and Sandusky Region* (Chicago: S. J. Clarke, 1929), 108.

3. Josephine Fassett, *History of Oregon and Jerusalem Townships, Lucas County, Ohio, 1837–1961* (Arkansas: Camden Co., 1961), 68.

4. Wilmot A. Ketcham, "Cedar Point in the Light of Olden Days," *Northwestern Ohio Historical Society Bulletin 9*, no. 1 (1937): 1–3.

5. Durant, Samuel W., *An Illustrated Historical Atlas of Lucas and part of Wood Counties, Ohio* (Chicago, Illinois: Andreas and Baskin, 1875).

6. "Magee Marsh Wildlife Area," publication no. 8 of the Ohio Department of Natural Resources, Division of Wildlife (1988), 2.

7. Ketcham, 1.

8. Edward Lamb, "The Grant of Lamb Beach to the Public" (Toledo, Ohio, privately printed, 1980), 1.

9. E. L. Fullmer, "The Toledo Cedar Point," *Ohio Journal of Science* 16 (1916): 216–18.

10. Fullmer, 216.

11. D. Handley, "How Many Ducks Nest in Ohio?" *Ohio Conservation Bulletin* 18, no. 4 (April, 1954): 18, 31, 32.

12. Noel Burns, *Erie, The Lake That Survived* (Totawa, New Jersey: Rowman and Allanheld, 1985), 141.

13. The Council included the Wolf Creek Sportsmen's Club, South Side Sportsmen's Club, Toledo Crow Club, Toledo Naturalists' Association, Fly Fishermen's Campfire, Men's Garden Club, and Lucas County Federated Women's Garden Clubs, plus interested individuals. Mrs. Neil Waterbury was president.

14. The marsh holdings acquired in 1961, with their owners and acreages:

Continental Marsh	Walter Apling	196.79 acres
(Unnamed)	Roy J. Dewey	16.03 acres
Douglas Marsh	B and T, Inc.	483.66 acres
Eisenhour Marsh	Carl Eisenhour et al.	235.25 acres
France Marsh	Enoch H. France,	
	Delphos Quarries, Inc.	289.27 acres
Hunter Marsh	Ira Hartman	80.0 acres
Ritter Marsh	George Ritter	573.61 acres
Searles Marsh	Ray Searles et al.	196.21 acres
Willow Point Marsh	Ruth Loughlin et al.	193.00 acres

15. Following Manke as Refuge managers were John Frye, 1971; James Carroll, 1974; Lee Herzberger, 1976; Michael Tansey, 1984; and the present manager, Charles Blair.

16. *Atlas of Ottawa County, Ohio* (H. J. Goodman, 1900).

17. Ibid.

18. Karl E. Bednarik, "Legends of Magee Marsh," *Ohio Conservation Bulletin* 11, no.1 (January, 1952): 6–8, 32.

19. Mark Shieldcastle, "Bald Eagles in Ohio—Today, Tomorrow," *Wild Ohio* (Spring, 1991), 6–7.

20. John G. Ketterer, "History of the Toussaint Shooting Club" (mimeographed report), 1969.

Chapter 3

1. Harold Mayfield, "The Changing Toledo region—a Naturalist's Point of View," *Northwest Ohio Quarterly* 34, no.2 (1962): 83–104, and references therein.

2. Ibid., 96–99.

3. Ibid., 97–98.

4. John Anderson, "Summer Birds of Winous Point in 1880, 1930, and 1960" (Mimeographed report) Winous Point Research Committee, 1960.

5. Laurel Van Camp, in *Toledo Naturalists' Association Annual* (1974), 11–17.

6. By Lynn Schuman, Sergej Postupalsky, and Laurel Van Camp.

7. E. L. Moseley, "Great Blue Herons in Sandusky County" (Mimeographed report), 1935.

8. Emery D. Potter, "Small Game of the Maumee Valley," *Toledo Commercial*, Mar. 19, 1870.

9. Quoted in A. J. Cook, *Birds of Michigan* (Lansing: Michigan Agricultural Experiment Station, 1893).

10. Quoted in Walter B. Barrows, *Michigan Bird Life* (Lansing: Michigan Agricultural College, 1912).

11. Anderson, 1960.

12. Quoted in Barrows, 1912.

13. J. M. Wheaton, "Report on the Birds of Ohio," in *Geological Survey of Ohio, Vol. 4: Zoology and Botany* (Columbus, Ohio: Nevins, 1882), 504.

14. Lynds Jones, *The Birds of Ohio* (Ohio State Academy of Science, Special Papers No. 6, 1903), 57.

15. Ibid., 29.

16. Anderson, 1960.

17. David B. Day, "Lake Erie and Michigan Marshes," *Transactions of the 20th North American Game Conference* 20 (1934), 29–35.

Appendix A

1. Harold Mayfield, "The Changing Toledo Region—a Naturalist's Point of View," *Northwest Ohio Quarterly* V. 34 (2), 1962, 92–97.

2. Emery D. Potter, "Small Game of the Maumee Valley," *Toledo Commercial*, March 19, 1870.

3. Charles S. Van Tassel, *History of the Maumee Valley, Toledo, and Sandusky Region* (Chicago: S. J. Clarke, 1929).

4. Karl Bednarik, "Muskrat in Ohio Lake Erie Marshes," Ohio Department of Natural Resources, Division of Wildlife, 1956.

5. Jack L. Gottschang, *A Guide to the Mammals of Ohio* (Columbus: Ohio State University Press, 1981).

6. Roger Conant, *The Reptiles of Ohio,* 2nd edition (Notre Dame, Indiana: The American Midland Naturalist, 1951).

7. Personal communication, May 9, 1972.

8. Charles F. Walker, "The Amphibians of Ohio, Part I: The Frogs and Toads," (Columbus: Ohio State Museum of Sci. Bulletin V. 1, No. 3, 1946).

9. Published by the Maryland Department of Natural Resources, Annapolis, Maryland and the U.S. Fish and Wildlife Service, Newton Corner, Massachusetts, 1988.

Bibliography

Anderson, John. "Summer Birds of Winous Point in 1880, 1930, and 1960." Winous Point Research Committee, 1960. (Mimeographed report.)

Andrews, Ralph. "A Study of Waterfowl Nesting on a Lake Erie Marsh." Masters Thesis, The Ohio State University, Columbus, Ohio, 1952.

Atlas of Ottawa County, Ohio. H. J. Goodman, 1900.

Barrows, Walter B. *Michigan Bird Life.* Lansing: Michigan Agricultural College, 1912.

Bednarik, Karl E. "Legends of Magee Marsh." *Ohio Conservation Bulletin* 11, no.1 (January, 1952): 6–8, 32.

Bednarik, Karl E. "Muskrat in Ohio Lake Erie Marshes." Columbus, Ohio Department of Natural Resources, Division of Wildlife, 1956.

Brown, Samuel. *Views of the Campaigns of the North-western Army, &c.* Philadelphia: William G. Murphey, 1815.

Burns, Noel. *Erie, The Lake That Survived.* Totawa, New Jersey: Rowman & Allanheld, 1985.

Conant, Roger. *The Reptiles of Ohio,* 2d ed. Notre Dame, Indiana: The American Midland Naturalist, 1951.

Cook, A. J. *Birds of Michigan.* Lansing: Michigan Agricultural Experiment Station, 1893.

Day, David B. "Lake Erie and Michigan Marshes." *Transactions of the North American Game Conference* 20 (1934), 29–35.

Durant, Samuel W. *An Illustrated Historical Atlas of Lucas and part of Wood Counties, Ohio.* Chicago, Illinois: Andreas and Baskin, 1875.

Fassett, Josephine. *History of Oregon and Jerusalem Townships, Lucas County, Ohio, 1837–1961.* Arkansas: Camden Co., 1961.

Forsyth, Jane L. "Late-glacial and postglacial history of western Lake Erie." *The Compass of Sigma Gamma Epsilon* 51, no.1 (1973): 16–26.

Fullmer, E. L. "The Toledo Cedar Point." *Ohio Journal of Science* 16 (1916): 216-18.

Goldthwait, Richard P. "Ice Over Ohio." In *Ohio's Natural Heritage,* edited by Michael B. Lafferty. Columbus: Ohio Academy of Science, 1979, 32-47.

Gottschang, Jack L. *A Guide to the Mammals of Ohio.* Columbus: The Ohio State University Press, 1981.

"Great Lakes Basin Commission Report," *The Communicator* (April, 1973).

Hamilton, Captain Frank E. "A Forgotten Port and Log Towing Revenue Cutters," *Telescope* (Great Lakes Maritime Institute, Detroit, Michigan) 18, no.1 (1969): 207-11.

Handley, D. "How Many Ducks Nest in Ohio?" *Ohio Conservation Bulletin* 18, no.4 (April, 1954): 18, 31, 32.

Hatcher, Harlan. *Lake Erie.* Indianapolis: Bobbs-Merrill, 1945.

Herdendorf, C. E., and Stuckey, R. L. "Lake Erie and the Islands." In *Ohio's Natural Heritage,* edited by Michael B. Lafferty. Columbus: Ohio Academy of Science, 1979, 214-235.

Herdendorf, C. E. "The Ecology of the Coastal Marshes of Western Lake Erie: A Community Profile." Biological Report 85. Washington, D.C.: Fish and Wildlife Service, 1987.

Herdendorf, C. E. "Paleogeology and Geomorphology." In *Lake Erie Estuarine Systems: Issues, Resources, Status, and Management.* U.S. Department of Commerce, National Oceanic and Atmospheric Administration, 1989, 35-70.

Herdendorf, C. E. and Bailey, M. L. "Evidence for an early delta of the Detroit River in western Lake Erie." *Ohio Journal of Science* 89, no.1 (1989): 16-22.

Historical Atlas of Ottawa County, Ohio. Chicago, Illinois: H. H. Hardesty, 1874.

Hough, Jack L. *Geology of the Great Lakes.* Urbana: University of Illinois Press, 1958.

Jaworski, E., Raphael, C. N., Mansfield, P. J. and Williamson, B. B. "Impact of Great Lakes water level fluctuations on coastal wetlands." In *Fish and Wildlife Resources of the Great Lakes Coastal Wetlands, Vol. 1: Overview,* edited by C. E. Herdendorf, S. M. Hartley, and M. D. Barnes. Washington, D.C.: U.S. Fish and Wildlife Service, 1981.

Jones, Lynds. *The Birds of Ohio.* Ohio State Academy of Science, Special Papers No. 6, 1903.

Ketcham, Wilmot A. "Cedar Point in the Light of Olden Days." *Northwestern Ohio Historical Society Bulletin* 9, no.1 (1937): 1-3.

Ketterer, John G. "History of the Toussaint Shooting Club," 1969. (Mimeographed report).

Lafferty, Michael B., ed. *Ohio's Natural Heritage*. Columbus: Ohio Academy of Science, 1979.

Lamb, Edward. "The Grant of Lamb Beach to the Public." Toledo, Ohio: privately printed, 1980.

Lewis, C. F. M. "Late quaternary history of lake levels in the Huron and Erie Basins." In *Proceedings of the 12th Conference on Great Lakes Research*. Ann Arbor, Michigan: International Association for Great Lakes Research, 1969.

Mayfield, Harold. "The Changing Toledo region—a Naturalist's Point of View." *Northwest Ohio Quarterly* 34, no.2 (1962): 83-104.

"Magee Marsh Wildlife Area." Publication no. 8 of the Ohio Department of Natural Resources, Division of Wildlife, 1988.

Morley, Thomas. "Uses and Values of Natural Areas." *Conservation Volunteer* 23, no. 135 (1960). Minnesota Department of Natural Resources.

Moseley, E. L. "Great Blue Herons in Sandusky County," 1935. (Mimeographed report.)

Potter, Emery D. "Small Game of the Maumee Valley." *Toledo Commercial*, Mar. 19, 1870.

Riddell, J. D. *Report No. 60 to the Governor and Legislature of Ohio*. Columbus, Ohio: James B. Gardiner, 1836.

Shieldcastle, Mark. "Bald Eagles in Ohio—Today, Tomorrow." *Wild Ohio* Spring (1991).

Thwaites, Reuben G. *The Jesuit Relations and Allied Documents, etc.*, Vol. 69. Cleveland, Ohio: Burrows Brothers Co., 1901.

Tiner, Ralph W., Jr. *Field Guide to Nontidal Wetland Identification*. Annapolis, Md: Maryland Department of Natural Resources and Newton Corner, Massachusetts: U.S. Fish and Wildlife Service, 1988.

Trautman, Milton B. *The Fishes of Ohio*. Columbus: Ohio State University Press, 1957.

Uhl, John B. *Official Atlas of Lucas County, Ohio*. Toledo, Ohio: Uhl Brothers, 1900.

U.S. Department of Commerce, NOAA, National Ocean Survey for 1972-73. *Monthly Bulletin of Lake Levels*. Detroit, Michigan: Lake Survey Center, 1973.

Van Camp, Laurel. In *Toledo Naturalists' Association Annual* (1974), 11-17.

Van Tassel, C. S. *History of the Maumee Valley, Toledo, and Sandusky Region*. Chicago: S. J. Clarke, 1929.

Walker, Charles F. "The Amphibians of Ohio, Part I: The Frogs and Toads." *Columbus, Ohio State Museum Science Bulletin* 1, no.3 (1946).

Wheaton, J. M. "Report on the Birds of Ohio." In *Geological Survey of Ohio, Vol. 4: Zoology and Botany*. Columbus, Ohio: Nevins, 1882.

Index

Note: numbers in bold face refer to figures; numbers followed by n refer to end notes or footnotes.

Amphibians in Lake Erie marshes (*see also* Frogs), 186–88

Aquatic plants: effect on, of erosion, 62, 107; effect on, of changing water levels, 93

Audubon Society, 81, 82, 156

Bienville, Céleron de (early explorer), 118

Bird banding, 64–67, 138–39, 158; as \aid to identification, 66; equipment, 65; reaction of birds to, 65

Bird calls, 46; as celebration of spring, 27–28; few, in August, 58; on a spring night, 34–38; rendering of, into words, 30–31, 49

Bird counts: aerial, (waterfowl) in Magee Marsh, 79, 129, 146, 149, 172; in Cedar Point Marsh, 129, 136; in Darby Marsh, 136, 139; in Ottawa National Wildlife Refuge, 81, 82, 136, 137, 156

Bird identification, 64–67; flycatchers, 30; owls, 83–84; shorebirds, 64, 175; warblers, 29, 64–67; waterfowl, 51–52, 64, 73

Bird records, keeping of, 137, 154–55 (*see also* Bird counts)

Birds: changing populations of, in Lake Erie marshes, 157–58, 166; numbers of, in winter, 81–82; numbers of, seen in Lake Erie marshes, 154–77

Birds of the Lake Erie marshes, 193–213

Bittern, 165; camouflage of, in cattails, 20, 50; nest of, in cattails, 2, 38; numbers of, in marshes, 166; voice of, 49

Black Swamp, 101, 104; draining and cutting of, 62, 104, 111

Blinds, duck, 69–70

Bono, village of (formerly Shepardsville), 106–07, 108

Breeding birds (*see* Nesting birds)

Brown, Samuel (early explorer), 100

Bullfrog, 187–88; diet of, 46, 54, 66; first appearance in spring, 25; trapping of, in Magee Marsh, 143; voice of, 28, 35, 46

Bullhead fishing, 25–26

Bunno, Frank (village of Bono named for), 107

Butterflies, 47, 58–60, 75; first appearance of, in spring, 19; monarch, migration of, 29, 58–60; monarch, tagged, 59–60

Byers, James (operated fishery at Crane Creek), 108

Cadillac, Antoine de la Mothe (early explorer), 118

Camp Perry, 108, **109**, 110

Canals: access to marshes via, 44, 112, 123; bullhead fishing in, 25–26;

filled by erosion, 44, 113, 126; for drainage, 107, 123

Cane: as marsh habitat, 2, 100; called "prairie" by early explorers, 100; deer in, 42; killed by high water, 144; response to changing water levels, 93, 100–101

Carp, effect of, on marshes, 47, 62

Carroll Township, 105, 107

Catawba Island, 166

Cattails: as marsh habitat, 2; birds that nest in, 38; killed by high water, 144; response to changing water levels, 93, 100–101

Cedar Creek: filled in to make Ward's Canal, 106, **110**; former bed of, in low-lying marshland, 110, 112

Cedar Point Marsh, 5, 44, 57, 106, 113, 118–29, **120–21**, 143, 162; bird counts conducted in, 129, 136; Bono family lived on beach of, 123–24; canals in, 126; commercial fishing on beach of, 107, 122, 124–25, 128; concentrations in, of songbirds, 28–29, 127; concentrations in, of eagles, 161; dikes rebuilt in, 123, 127, 129; donated to North American Wildlife Foundation, 128; eagles' nests in, 159, 160; effect of drought on, 100–101, 123; flooded, 101, 123; hunting duck in, 71, 75; hunting pheasant in, 123, 126, 128; night herons nested in, 164; outer beach of, eroded, 122, 127; President Eisenhower hunting in, 75, 128; pumping stations in, 123, 126–27; shorebirds in, 176; special vulnerability of, to flooding, 127; trapping of snapping turtles in, 186; undiked portion of, flooded, 116, 129; bird counts in, 129

Cedar Point National Wildlife Refuge, 118, 128 (*see* Cedar Point Marsh)

Cedar Point Shooting Club, 112, 119, 142; clubhouse of, 122, 129; lawsuit by, against fishery, 122, 124–25

Chick Islands, 166

Christmas Bird Census, Audubon Society, 81, 82, 156, 169

Click beetle, 61

Clothing, protective, 81

Concentrations (*see* Songbirds; migration; Little Cedar Point; Navarre Marsh; Magee Marsh; Shorebirds, Eagles)

Continental Marsh, 133

Cooley Canal (Reno Side Cut Ditch), 106, 112, 118; present boundary of Cedar Point Marsh, 122

Coot, American (mudhen): courting, 23

Cormorant, double-crested 18, 76, 166–67; nests on West Sister Island, 167

Courtship: of American bittern, 49; of American coot, 23; of harrier (marsh hawk), 23; of waterfowl, 21–23

Cowbird, brown-headed, 38

Crane Creek, 31, 100, 133, 167, 170; commercial fishing at mouth of, 108; eagles congregate at mouth of, 146; straightened, 111

Crane Creek Shooting Club, 141

Crane Creek State Park 61, 140, 147, 160, 168

Crane Creek Wildlife Experiment Station (at Magee Marsh), 65, 79, 82, 143–46; aerial waterfowl counts, 79, 129, 146; Canada goose project, 144–46, 173–74; eagle project, 146

Darby Marsh (part of Ottawa National Wildlife Refuge): bird counts in, 136, 139, 168; exchanged for Navarre Marsh, 134, 139, 149

Davis-Besse Marsh (*see* Navarre Marsh)

Davis-Besse nuclear power plant, 29, 43, 65, 110, 136, 149

Decoys, 71–72, 73

Deer, white-tailed 3, 42–43, 180; damage by, to banding nets, 67

Detroit River, 16, 64, 96

Dikes: access to marshes via, 17, 33, 34, 44, 112, 123; as marsh habitat, 3, 113; as nurseries, 39–42; breached by floods, 9, 101, 114, 115, 127, 129, 148; definition of, 111; dens in, 3, 41, 42; effect of, on fish, 25, 46–47, 113; effect of, on wild rice, 113; first construction of, along Lake Erie, 111–12; stone, expense of, 114, 116,

129; stone, protection offered by, 9, 96, 99, 100, 115–16; vegetation on, 3, 33, 43, 44, 54, 68, 77–78, 112; woodchuck dens in, 40

Dogs, hunting, 72–73

Dragonfly, 47, 52–54, 75; diet of, 53–54; distinguished from damselfly, 53; reproduction of, 53; spring migration of, 19; survival of, 52–53, 54

Duck blinds, 69, 144 (see also Hunting, waterfowl)

Duck calling, 72

Ducks (see also Waterfowl): courtship of, 22; enemies of, 46; hunting of (see Hunting, waterfowl); molting of, in summer, 50–51

Ducks, diving: concentrations of, 76, 82; courtship, 22; passion of, for zebra mussels, 62; spring migration of, 15

Eagle, bald, 158–62; concentrations of, 161–62; diet of, 82, 159; effect of pesticides on eggs of, 50, 154, 159; fostering program for, 159–60; migration of, 146, 161; nests of, 17, 158–61; nests of, in great blue heronries, 161; project, at Crane Creek Wildlife Experiment Station (Magee Marsh), 146; reaction of marsh birds to, 56

Egret, cattle, 32, 165, 166

Egret, great, 163–64; history of, in Lake Erie marshes, 157, 163; in fall marsh, 78; nests on West Sister Island, 2, 39, 157, 163; swallowing fish, 46

Egret, snowy, 57, 165, 166

Egrets, 18, 163, 165

Eisenhower, Dwight David: hunting in Cedar Point Marsh, 75, 128

Erie County, number of eagle's nests in, 159

Erie Industrial Park, 108, 109

Erie Marsh (Michigan), 5; herons nesting in or near, 163, 164

Erie Proving Ground, 150

Erie Township, 105, 139

Erosion: canals filled in by, 44, 113; dikes protected against, by planting willows, 112; effect of, on water

quality in Maumee and Sandusky Bays, 62, 107; in Great Lakes watershed, 98

Falcon, peregrine, 18; reaction of marsh birds to, 56–57

Farming, in drained swamp forest soils, 107

Fish: bullheads, 25–26; carp (see Carp); in marshes, 25, 46–47, 113

Fishing: at Magee Marsh, 5, 140; at Metzger Marsh, 26, 147–48; commercial, 62, 63, 107–8; 122, 124–25; on Ward's Canal, 26

Flycatchers: arrival in spring, 24; distinguishing among, 30–31

Flyways, Atlantic and Mississippi, 15, 155 (see also Migration)

Food supply, for wildlife, in winter, 10–11

Foxes, 41, 179

Fox, red: dens of, 3, 41; diet of, 41; pelts, sale of, 179; changing population of, 179

Frogs, 2, 45, 46; first appearance of, in spring, 19, 21, 25; gradual disappearance of, in fall, 68; last varieties of, in fall, 75; trapping of, in Magee Marsh, 143; voices of, in night marsh, 35; voices of, in spring, 19, 21, 28; voices of, in summer, 46

Fur sales, 143, 179, 180

Geese: courtship of, 21; fall migration of, 68, 76, 174; numbers of, in fall marshes, 79, 82, 174; spring migration of, 15, 17–18

Glaciation of Ohio, 90–92

Glacier: "retreat" of, 90–91; moraine left by, 90–91, 91; weight of, 90, 92, 95

Goose: hunting (see Hunting, waterfowl); project, Crane Creek Wildlife Experiment Station (Magee Marsh), 145, 173–74; roundup, Magee Marsh and Ottawa Refuge, 174

Grass meadow: as marsh habitat, 2

Great Lakes, upper: influence of, on Lake Erie water levels, 96, 98

Great Lakes watershed: population

boom in, 98
Grebe, pied-billed, 35, 48
Gull, 167–70
Gull, herring, 169
Gull, laughing, 169
Gull, ring-billed, 167–68; effect on, of
    1988 heat wave, 168; nests in
    Maumee Bay, 167–68

Hankison Marsh (*see* Sand Beach
    Marsh)
Harrier (marsh hawk): mating flight of,
    23
Hawks: changes in populations of, near
    Lake Erie, 158; counting and band-
    ing of, 158; diet of, 82; killing of, 24;
    northward migration of, 16, 24, 158;
    numbers of, in winter, 13, 82–83;
    southward migration of, 67, 76, 79
Heron, black-crowned night, nests: on
    Little Cedar Point, 128, 164; near
    Erie Marsh, 164; at Magee Marsh,
    143, 164; on West Sister Island, 2,
    35, 39, 164
Heron, great blue, 162–63; call of, when
    disturbed, 37–38; heronry of, eagles
    nesting in, 161; heronry of, on West
    Sister Island, 2, 39, 155; once nested
    at Magee Marsh, 143, 163
Heron, green-backed, 34, 38, 84, 164–65
Heron, little blue, 32, 57, 165, 166
Heron, tricolored, 32, 57, 165
Heron, yellow-crowned night, 32, 165,
    166
Heronry, 162–63 (*see also* West Sister
    Island and entries under Heron)
Herons, 18, 162–66; in fall marsh, 67,
    78; in summer marsh, 57; largest
    recorded number of, 162; spring
    arrival of, 18, 32
Horizon Enterprises, 132–33, 140
Housing developments: on drained
    marshlands, 108, 112, 117 (*see also*
    Reno Beach, Howard Farms); on
    higher ground, 108, 112
Howard Farms, 110, 112; flooded, 114,
    115, 123
Hunters: role of, in preserving
    marshlands, 5, 69, 130
Hunting: in Cedar Point Marsh, 71, 75,
    123, 126, 128; in Magee Marsh,

144–46; in Toussaint Marsh, 73,
    150–52; no threat to wildlife, 5, 69,
    144, 150; of upland game birds, 80,
    126, 150–52
Hunting, waterfowl, 68–75, 79, 174;
    from a blind, 69–70, 71; from a boat,
    70–71, 74–75; from a ditch, 71; in
    Ottawa National Wildlife Refuge,
    69, 79, 136, 144; with a dog, 72–73;
    with decoys, 71–72, 73, 74
Hunting season, waterfowl: dates of,
    69; numbers of ducks vary during,
    71

Ibis, glossy, 32, 57, 166
Ibis, white-faced, 32, 166
Ice Age, 89–93

Jerusalem Township, 105, 107

Killdeer: handicap to birders, 56;
    nesting, 20, 175; voice of, 16, 35

Lacarpe Creek, 111–12
Lacarpe Shooting Club, 139
Lake Erie: barrier to migrating birds,
    16, 24, 28; basin, effect of glacier on,
    90, 95; basin, shape and orientation
    of, 89, 96–97; birds in and near, in
    fall, 76, 82; fishing in, commercial,
    62–63, 107–08, 122, 124–25; fishing
    in, sport, 107; monarch butterflies
    crossing, 59; seiches on, 97–98;
    storms on, 96, 97, 98, 99, 115; zebra
    mussels in, 61–62, 63
Lake Erie water level: average, 9, 98, 99;
    changes in, 92–93, 96–99, 113–16;
    effect of upper Great Lakes on, 96,
    98
Lake Erie Marshes (*see* Marshes, Lake
    Erie)
Lake Maumee (ancestor of Lake Erie),
    91–93, 94
Lake Shore and Michigan Southern
    Railroad, 139
Lamb Beach, 120–21, 122, 160
Lamb, Clarence (founded fishery on
    Lamb Beach), 122, 124–25
Lamb, Edward (son of Clarence Lamb),
    122, 124–25
Léry, Chaussegros de (early explorer),

118

Little Cedar Point, 108, 120–21; formerly songbird concentrations on, 28, 127; history of, 118–19; lighthouse on, 119; loss of trees on, 28, 65; prothonotary warbler on, 31; tavern on, 119

Little Cedar Point sandbar: black-crowned night herons nesting on, 128; changing position of, 120, 127; piping plover nesting on, 128, 175; terns nesting on, 128, 169–70

Locust Point: subdivision, 108; two early settlements called, 106

Long Beach (subdivision), 108

Lucas County, 105; number of eagle's nests in, 159

Lumbering, in area swamp forest, 106–07

Magee, John Nicholas, 142–43

Magee Marsh, 31–32, 106, 140–47; accessibility of, to public, 140; aerial waterfowl counts conducted by officials of, 79, 129, 146, 172; bird trail at, 5–6, 53, 140, 146–47; Crane Creek Wildlife Experiment Station at (see Crane Creek Wildlife Experiment Station); early history of, 140–43; fishing site at, 5, 140; herons nested in, 143, 163, 164; hunting in, 69, 70, 140, 142, 144–45; management of, 142–43, 144; prothonotary warbler in, 32, 143; remnants of swamp forest in, 104; shorebirds in, 176; songbird concentrations on, 29, 127, 140; Sportsmen Migratory Bird Center at, 140; trapping in, 140, 143; waterfowl production project, 173

Magee Marsh Wildlife Area, 140, 144

Mallard Club Marsh (see Maumee Bay State Park)

Mammals in Lake Erie marshes, 179–84

Manke, Alfred (first manager of Ottawa National Wildlife Refuge), 133

Marinas, 110

Marsh: as source of genetic diversity, 4; definition of, l; effect of changing water levels in, 93, 96, 99–101, 102–03, 144; effect on birds of

changes in habitats associated with, 157–58; habitats, 1–5, 99–104, 113, 129 (see also Dike, Thicket, Pond, Swamp forest, Sand ridge, Cattail, Cane, Grass meadow); human recreational activities in, 5; in winter, silence and solitude of, 81–82, 84–85; plants associated with, 189–92; sounds of, in spring, 27–28, 34–38; visit, at night, 34–38; visit, in June, 43–50; management of, 111, 112–14, 133, 139, 142, 144

Marsh wren: nest of, 2, 48–49; voice of, 36, 48

Marshes, Lake Erie: accessibility of, 5, 44, 111, 140; amphibians in, 186–88; average date frozen over, 78; birds of, 193–213; buffer between lake and swamp forest, 93, 117; changes in bird populations of, 157–58; controlling water levels in, 111, 112–16; creation of, by glaciation, 93–101; creation of, by hunters, 113, 133; drainage of, by farmers and developers, 111–12, 117, 142; effect of changing water levels on, 99–101, 117; effect of drought on, 100–101, 113; effect of high lake levels on, 9, 96, 114–16; government agencies and, 6, 114–16, 117, 128–29, 149; low-lying, west of Turtle Creek, 111, 112, 115, 117, 123, 118–49, 152; mammals of, 179–84; most primitive lands in Ohio, 1–2; numbers of bird varieties that nest in, 39, 157; plant species in, 189–92; reptiles in, 184–86; settlements near, 105–08 (see Settlements); shelter for migrating birds, 2, 15, 155–57 (see Flyways); small, privately owned, 152–53; fate of undiked portions of, 10, 116, 148; fishes in undiked portions of, 25, 26–27

Maumee Bay: cormorants at, 166; diving ducks in, 76; eagles' nest near, 159; gulls' nest in, 167–68, 169; marshes in, 93; silt in, from draining of Black Swamp, 62, 107; terns' nest in, 170–71

Maumee Bay State Park, 148–49

Maumee River, 100

Mayflies, 63

Metzger Farms, 112, 141, 148

Metzger Marsh, 147-48, 168; fishing in, 26, 147-48; hunting in, 70; terns nesting in, 169; undiked, effect of high water on, 116, 148

Migrating birds, reluctance of, to cross Lake Erie, 16, 24, 155

Migration: bird, revealed by banding, 66; eagle, 146, 161; of hawks and crows, northward, 16, 18, 157; of hawks and crows, southward, 67, 76, 79, 157; of herons, northward, 18, 32; of monarch butterflies, northward, 29; of monarch butterflies, southward, 59-60; of owls, northward, 16; of shorebirds, northward, 19-20, 24, 32, 37; of shorebirds, southward, 52, 55, 67; of songbirds (see also Warblers), northward, 16-18, 20, 25, 29, 30, 37; of songbirds, southward, 78, 79; of waterfowl, northward, 15-18, 157; of waterfowl, southward, 57, 68, 76, 78, 157; reverse, 28; spring, dragonflies, 19; spring, fish, 25

Migratory Bird Conservation Commission, 133

Mink: "perfect predator," 40; tracks in snow, 12; trapping of, 80

Molting: of ducks, 50-51; of shorebirds, 52

Mominee, Cornelius (guide, Cedar Point Club), 75, 128

Mominee, Joseph (founder of Momineetown), 119

Morley, Thomas (wildlife biologist), 4

Mouse, white-footed, 11-12

Muddy Creek (Sandusky County), 161

Mudhen (see Coot)

Muskrat: benefits from marsh management, 42, 80, 111, 112-13, 136; dens of, in dikes, 3, 42; diet of, 42; enemies of, 42; in marsh ponds, 2, 42, 45; tracks, in snow, 10; trapping of, 42, 80, 136, 143

Navarre Marsh, 32, 136-39; banding in, 65, 138-39; exchanged for Darby as site for power plant, 134-36, 139, 149; managed by Ottawa National Wildlife Refuge, 65, 134-36; protection of, by limestone ridge, 135, 136; remnants of swamp forest in, 104; songbird concentrations on, 29, 127, 138; variety of habitats in, 138

Navarre, Peter ("Peter the Scout"), 119

Navarre Shooting Club, 136

Nest (see also Heronry): bird's, used by white-footed mouse, 11; of eagle, 17, 158-61; of killdeer, 20; of larger birds, 84; of prairie horned lark, 17; of prothonotary warbler, 31-32; of snapping turtle, 48; of vole and shrew, 26

Nesting birds, 17, 20, 38-39 (see also entries under individual species, e.g., Herons, Waterfowl); benefit from stabilized marsh water levels, 113; in cattails, 38; in open fields, 38; in swamp forest, 38-39; in thickets, 38; numbers of, in Lake Erie Marshes, 157, 158-77; on West Sister Island, 39

Nesting boxes: for prothonotary warblers, 31-32; tree swallows in, 32; wood duck, 45; wood duck, used by raccoons, 40

Nesting platform, for eagles, 161

Norman's Strait (breach in Cedar Point Marsh beach), 127

North Cape, Michigan: eagles' nest at, 161; heronries at, 163, 164

Oak Openings, 84, 101

Ohio Department of Natural Resources, 143, 159

Ohio Division of Parks and Recreation, 140

Ohio Division of Wildlife, 59, 140, 148, 166, 173

Ohio-Michigan boundary dispute, 106

Opossum, 41-42; "playing possum," 41; diet of, 42; effect of, on waterfowl production, 113

Ottawa County, 105, 139; number of eagles' nests in, 159

Ottawa National Wildlife Refuge, 5, 60, 82, 106, 108, 128, 129-40; additions to, 129-30, 134-36; bird counts in, 81, 82, 136, 137, 156; birth of, 130-33; concentration of eagles in,

162; cormorants nesting in, 166–67; eagles' nest in, 159; flooded area of, 133, 170; hiking trails in, 29, 136; hunting waterfowl in, 69, 79, 136; insects in, 60–61; management of, 133, 134–35; management of Navarre Marsh by, 65, 135; primary goal protection of wildlife, 134–35; private marshes comprising, 130, 131, 133, 133n; prothonotary warblers in, 31; remnants of swamp forest in, 104; shorebirds in, 176; trapping in, 136; waterfowl in, 57, 82

Ottawa River, 161

Owl pellets, 84

Owls: as predators, 14, 37; counting of, 158; identification of, 83–84; migration of, 16; voices of, 84

Pheasant hunting, 80; in Cedar Point Marsh, 123, 126, 128

Pheasants, 157

Pintail Farms, 112, 114, 139

Pintail Marsh, 72, 139–40, 148; now part of Ottawa National Wildlife Refuge, 132, 133

Plover, golden, 19–20

Plover, piping, 128, 175

Poachers: in Magee Marsh, 143; marsh owners wary of, 5, 123, 124

Pond: as marsh habitat, 2

Portage River, 100

Potter's Pond, in Cedar Point Marsh, 126

Prairie horned lark, 17

Predation, 14, 48

Preying mantis, 60

Prohibition, 126

Pumps, water: creation of marshes using, 133; in marsh management, 112, 113–14, 123, 126–27, 139

Punt, 44, 126

Quail: hunting of, 80, 152; now gone from marshes, 152, 157

Raccoons: captured for pets, 40–41; diet of, 41; effect of, on waterfowl production, 41, 113; with spring fever, 21; nests of, 26, 40; trapping of, 136

Raisin River: massacre at, 119; yellow-crowned night heron nest along, 166

Reno Beach (subdivision), 110, 112; flooded, 114, 115, 123

Reptiles in Lake Erie marshes, 184–86

Riddell, John D. (early explorer), 100

Rocky Ridge Shooting Club, 141–42

Rocky Ridge, village of, 142

Rusha Creek, 150

Sand Beach (subdivision), 108, 112

Sand Beach Marsh, 110, 135, 136, 149

Sand ridges: as marsh habitat, 3–4; as natural barriers, 3–4, 96, 97; destroyed by stone dikes, 4

Sandusky Bay: cattle egrets nest on, 165; cormorants seen on, 167; diving ducks in, 76; eagles congregate near, 146; marshes in, 93

Sandusky Bay Marshes, 39

Sandusky County: eagles' nests in, 159, 161; herons' nests in, 163

Seiche, 97–98

Settlements: French, 105–6, 108; Indian, 105; on high ground, 108 (see also Bono, Toussaint, Housing developments)

Shorebirds, 175–77; concentrations of, 55, 177; counts of, 176; fall migration of, 52, 55, 67, 76, 175; flocking instinct of, 56; identification of, 64, 175; nesting in marshes, 175; spring migration of, 19–20, 24, 32, 37, 175; voices of (see Bird calls)

Shrews: ferocity of, 12–13; nests of, 26; runways of, 79

Skink, five-lined, 3, 33, 184

Skunk: diet of, 40; effect of, on water-fowl production, 113; tracks in snow, 12

Smith, James (described 1757 hunt on Little Cedar Point), 118–19

Smuggling, of beer during Prohibition, 126

Snake, fox, 2, 48, 184

Snakes, 184–86; first appearances in spring, 19, 21; last appearance in fall, 75

Snapping turtle (see Turtle, snapping)

Songbirds (*see also* Bird banding, Bird identification, Warblers): concentrations, 28–29, 127, 138; in winter, 79, 81, 82; nesting, 38; northward migration of, 16–18, 20, 30, 37; southward migration of, 78; voices of (*see* Bird calls)
South Bass Island, 166
Sportsmen Migratory Bird Center (at Magee Marsh), 140
Stifel, Doris, 59–61
Swallows, 75: first, in spring, 24; flocking in August, 58
Swamp forest: as marsh habitat, 3; bordering Lake Erie Marshes, 101, 104 (*see also* Black Swamp); farming in drained soils of, 107; in drained Lake Erie basin, 92, 93; killed by high water, 93, 100; lumbering in, 106; nesting birds in, 38
Swans, 15, 17–18

Tern, common: migrating, 170; nesting, 47, 128, 154, 169–70
Tern, Forster's, 170–71
Terns, 169–71
Thickets: as marsh habitat, 3, 38, 77; between swamp forest and marsh, 101; color of, in fall, 68, 77; nesting birds in, 38
Thompson, Eugene (cofounded fishery on Lamb Beach), 122
Toledo Edison Company, 65, 134, 161, 168
Toledo Naturalists' Association, 177
Toussaint Marsh, 149–52; effect of high water on vegetation in, 101, 102–03, 150; flooded, 115; hunting in, 73, 150–52; prairie grass (cane) in, 100; remnants of swamp forest in, 104
Toussaint River, 93, 100, 109, 150, 152, 166; diked, 111–12; early route through marshes, 105–06
Toussaint Shooting Club, 73, 150–52, 171
Toussaint, village of (formerly Locust Point), 106
Tracks: in mud, 48; in snow, 11–12
Trappers, French, 105
Trapping: in Ottawa National Wildlife Refuge, 136; no threat to wildlife,

4–5, 80; of mink, 80; of muskrat, 42, 80, 143; of raccoons, 41, 80; of snapping turtles, 41, 143
Turtle snapping: effect of, on waterfowl production, 41, 113; nest of, 48; trapping of, in Cedar Point Marsh, 186; trapping of, in Magee Marsh, 143
Turtle Creek, 93, 100, 104, **109**, 112, 140, 146, 152; commercial fishing at mouth of, 108; diked, 111–12; public fishing access on, 26, 140
Turtles, 2, 3, 19, 21, 36, 45, 186–87; first appearance of, in spring, 25; last of, in fall, 75

U. S. Bird Banding Laboratory, 65
U. S. Fish and Wildlife Service, 5, 114, 123, 129–30, 136, 159

Voles, 11, 13, 26, 82

Warbler, prothonotary, 31–32, 143
Warblers (*see also* Bird banding): concentrations of, on bird trail at Magee Marsh, 29; difficulty of identifying in fall, 64; in fall, 75; in winter, 82; spring migration of, 25, 29, 30, 37
Ward, Eber Brock, 106, 107, 108
Ward's Canal, 108, 112, 148; Cedar Creek diverted into, 106, 110; commercial fishing at mouth of, 107; construction of, 106; fishing on, 26; formed original border of Cedar Point Marsh, 119
Waterfowl, 171–74; aerial counts of, 79, 129, 146, 149, 172; breeding populations, estimated, 171–73; courtship, 21–23; eclipse plumage of, 51, 64; habitat preferred by nesting, 113; hunting of (*see also* Hunting, waterfowl), 68–75, 79, 136, 144–45, 174; northward migration of, 15, 17, 20; numbers of, in fall marsh, 79, 150; numbers of, in winter marsh, 82; production, 111, 113, 171, 173–74; production, effect of snapping turtles and raccoons on, 41; southward migration of, 57, 68, 76; voices of, 35–36, 76
West Sister Island: cormorants nesting

on, 167; eagles nesting on, 159, 161; egrets nesting on, 163, 165; heronries on, 2, 35, 39, 155, 162, 163, 164, 165; herring gulls nesting on, 169; national wildlife refuge on, 130

Whirligig beetles, 60

Wildflowers: first, in spring, 20, 25; in and near ponds, 43, 47–48, 54–55; in forest and thicket, 33; summer, 54–55

Willow Beach (subdivision), 108

Winous Point Marsh (Sandusky Bay): great egret nested near, 163–64; heronry in, 161, 162, 163; record of bird species in, 158

Wolf Creek, 126

Woodchuck: dens of, 3, 39–40; in platinum blond fur, 68; with spring fever, 21

Wyandot County, 159

Zebra mussels, 61–62, 63

CPSIA information can be obtained
at www.ICGtesting.com
Printed in the USA
LVHW080604270921
698763LV00006B/95